Principles and Standards for School Mathematics Navigations Series

NAVIGATING *through* GEOMETRY *in* GRADES 3–5

M. Katherine Gavin
Louise P. Belkin
Ann Marie Spinelli
Judy St. Marie

Gilbert J. Cuevas
Grades 3–5 Editor
Peggy A. House
Navigations Series Editor

NATIONAL COUNCIL OF
TEACHERS OF MATHEMATICS

1906 Association Drive, Reston, VA 20191-1502
(703) 620-9840; (800) 235-7566; www.nctm.org

Fourth printing 2006

Library of Congress Cataloging-in-Publication Data

Navigating through geometry in grades 3-5 / M. Katherine Gavin … [et al.].
p. cm. -- (Principles and standards for school mathematics navigations series)
Includes bibliographical references.
ISBN 0-87353-512-X
1. Geometry--Study and teaching (Elementary)--United States. I. Gavin, M. Katherine. II. Series.

QA461 .N314 2001
372.7--dc21

2001054366

The National Council of Teachers of Mathematics is a public voice of mathematics education, providing vision, leadership, and professional development to support teachers in ensuring mathematics learning of the highest quality for all students.

Printed in the United States of America

Table of Contents

About This Book

> Geometry is grasping space ... that space in which the child lives, breathes and moves. The space that the child must learn to know, explore, conquer, in order to live, breathe, and move better in it.
>
> —Freudenthal, *Mathematics as an Educational Task*

Welcome to *Navigating through Geometry in Grades 3–5!* This book is about the "big ideas" of geometry presented in *Principles and Standards for School Mathematics*, by the National Council of Teachers of Mathematics (NCTM) (2000). These four ideas are shape, location, transformations, and spatial visualization. These quintessential concepts in school geometry are summarized in the introduction to the book and related specifically to grades 3–5 at the start of each chapter. Please take time to read the introductions to both the book and the chapter before you begin the activities in each chapter. Doing so will help you understand the various connections to geometry in grades 3–5 discussed in *Principles and Standards for School Mathematics* and deepen your understanding of the big ideas and the important mathematics that is behind each one.

The activities in this book are designed with you, the teacher, in mind. Although the activities for each big idea are not intended to constitute a complete unit, the activities are sequential, one building on another. Also, many connections among the big ideas have been emphasized throughout. The final activity, Geo City, incorporates the major concepts presented throughout the book. Designed as a project, this activity is effective as an interactive, performance-based assessment.

Each chapter begins with a brief overview of the concepts included in the big idea. The activities, which contribute to the development of those concepts, follow the same format throughout the book: Goals for students' learning are presented and are connected to the expectations for students in grades 3–5 outlined in the geometry section in *Principles and Standards for School Mathematics*. The prior knowledge expected of students, the materials necessary for conducting the activity, and the learning environment are then discussed, and the important geometric terms emphasized in the activity are listed. Since language development is very important in these grades, definitions for terms that may not be familiar to teachers have been included in some instances. Research has shown that imprecise language is very evident in students' work in geometry (Burger and Shaughnessy 1986). Correct language is essential for the development of conceptual understanding in geometry. Teachers are encouraged to introduce the important terms to students during the activity and to make sure that the students understand them before moving on. Using the terms frequently throughout the year will reinforce students' understanding and correct use of the terms.

The "Engage" section of each activity sets the stage for the exploration. This section, which is sometimes a miniexploration, is an important part of the lesson; its success rests on the teacher's efforts to spark

Key to Icons

Principles and Standards

CD-ROM

Blackline Master

Three different icons appear in the book, as shown in the key. One alerts readers to material quoted from *Principles and Standards for School Mathematics,* another points them to supplementary materials on the CD-ROM that accompanies the book, and a third signals the blackline masters and indicates their locations in the appendix.

interest, give clear directions, and sometimes provide initial instruction. The key is to engage without overinstructing, which can be difficult. In our desire for our students to succeed, we teachers often "help" them too much. We must remember that learning is much more powerful when it is gleaned through discovery—especially when some struggle is involved—than when it is imparted through telling.

In "Explore," the activities are described and suggestions are given for guiding the students' explorations. The blackline masters that accompany some of the activities are intended to help you organize the activities. The masters are signaled by an icon and can be found in the appendix, along with solutions to the problems. They can also be printed from the CD-ROM that accompanies the book. The CD, also signaled by an icon, contains applets for students to manipulate and resources for professional development.

The "Assess" section offers suggestions for ongoing informal and formal assessment. Please be aware that the time requirements for the activities differ. Some will take one class period of about one hour. Others, however, may take several days to complete, with the "Engage" section introduced on the first day, followed by a couple of days for the exploration and a final day for summary and assessment.

The "Extend" section can guide you in providing challenges that require high-level thinking for students who are ready to extend their learning. The tasks suggested in this section can be used in a learning center, in individualized instruction, or in a flexible-grouping program to give students differentiated instruction. "Where to Go Next in Instruction?" will help you see where an activity fits into the curriculum and what might follow logically to further develop the concepts investigated in it.

The authors hope you find that the activities in this book enrich your curriculum and help your students develop a strong sense of geometric concepts and relationships—help them "grasp the space in which they live." Foremost, we hope that in your explorations, you and your students experience the joy and wonder of geometry and other mathematics.

GRADES 3–5

NAVIGATING *through* GEOMETRY

Introduction

Both in the development of mathematics by ancient civilizations and in the intellectual development of individual human beings, the spatial and geometric properties of the physical environment are among the first mathematical ideas to emerge. Geometry enables us to describe, analyze, and understand our physical world, so there is little wonder that it holds a central place in mathematics or that it should be a focus throughout the school mathematics curriculum.

When very young children begin school, they already possess many rudimentary concepts of shape and space that form the foundation for the geometric knowledge and spatial reasoning that should develop throughout the years. *Principles and Standards for School Mathematics* (National Council of Teachers of Mathematics [NCTM] 2000) recognizes the importance of a strong focus on geometry throughout the entire prekindergarten–grade 12 curriculum, a focus that emphasizes learning to—

- analyze characteristics and properties of two- and three-dimensional geometric shapes and develop mathematical arguments about geometric relationships;
- specify locations and describe spatial relationships using coordinate geometry and other representational systems;
- apply transformations and use symmetry to analyze mathematical situations;
- use visualization, spatial reasoning, and geometric modeling to solve problems. (P. 41)

Geometry not only provides a means for describing, analyzing, and understanding structures in the world around us but also introduces an experience of mathematics that complements and supports the study of other aspects of mathematics such as number and measurement. Geometry offers powerful tools for representing and solving problems in all areas of mathematics, in other school subjects, and in everyday applications. *Principles and Standards* presents a vision of how geometric concepts and reasoning should develop and deepen over the course of the school mathematics curriculum. The *Navigating through Geometry* books elaborate that vision by showing how important geometric concepts can be introduced, how they grow, what to expect of students during and at the end of each grade band, how to assess what students know, and how representative instructional activities can help translate the vision of *Principles and Standards* into classroom practice and student learning.

Foundational Components of Geometric Thinking

The Geometry Standard emphasizes as major unifying ideas *shape* and the ability to analyze characteristics and properties of two- and three-dimensional objects and develop mathematical arguments about geometric relationships; *location* and the ability to specify positions and describe spatial relationships using various representational systems; *transformations* and the ability to apply motions, symmetry, and scaling to analyze mathematical situations; and *visualization* and the ability to create and manipulate mental images and apply spatial reasoning and geometric modeling to solve problems. Each of these components of geometric thinking requires nurturing and developing throughout the school curriculum.

Analyzing characteristics and properties of shapes

By the time the youngest children begin formal schooling, they have already formed many concepts of shape, although their understanding is largely at the level of recognizing shapes by their general appearance and they frequently describe shapes in terms of familiar objects such as a box or a ball. In the primary grades, children should have ample opportunities to refine and focus their understanding and to gradually develop a mathematical vocabulary. They also should learn to recognize and name the parts of two- and three-dimensional shapes, such as the sides and the "corners," or vertices. Teachers should provide frequent hands-on experiences with materials, including technology, that help the students focus on attributes of various shapes, such as that a square is a special rectangle with all four sides the same length or that pyramids always have triangular faces that meet at a common point. Experiences that promote such outcomes include building and drawing shapes; comparing shapes and describing how they are alike and how they are different; sorting shapes according to one or more attributes; cutting or separating shapes into component parts and reassembling the parts to form the original or different shapes; and identifying shapes found in everyday objects or in the classroom, home, or neighborhood. Throughout such activities, teachers must take care to ensure that the children encounter both examples and nonexamples of common shapes and that they see those examples in many different contexts and orientations so that they learn to identify a triangle or a rectangle, for example, no matter what material it is made of or how it is positioned in space.

As children progress to the higher elementary grades, they should continue to identify, compare, classify, and analyze increasingly more complex two- and three-dimensional shapes, and they should expand their mathematical vocabulary and refine their ability to describe shapes and their attributes. As they do so, they begin to develop generalizations about classes of shapes, such as prisms or parallelograms, and to formulate definitions for those classes. They also include in their study not only two- and three-dimensional shapes but points, lines, angles, and more-precise relationships such as parallelism and perpendicularity. They begin to explore properties of area and perimeter and to pose questions related to those measurement concepts; they might, for example, use tangram pieces to investigate whether shapes that are different, such as a rectangle, a trapezoid, and a nonrectangular parallelogram, can have the same area. They also develop and explore concepts of congruence and similarity, which they express in terms of shapes that "match exactly" (congruence) or shapes that "look alike" except for "magnifying" or "shrinking" (similarity). In grades 3–5, there should be a growing emphasis on making conjectures about geometric properties and relationships and formulating mathematical arguments to substantiate or refute those conjectures; for example, students might use tiles or grid paper to show that whenever the sides of one square are twice as long as the sides of another square, then four of the smaller squares will "fit inside" or "cover" the larger square, or they might measure to demonstrate that a rectangle, trapezoid, and nonrectangular parallelogram that have equal area do not necessarily have the same perimeter.

The informal knowledge and intuitive notions developed in the elementary grades receive more-careful examination and more-precise description in the middle grades. Descriptions, definitions, and classification schemes take account of multiple properties, such as lengths, perimeters, areas, volumes, and angle measures, and students should use those characteristics to analyze more-sophisticated relationships by, for instance, developing a classification scheme for quadrilaterals that accurately represents some classes of quadrilaterals (e.g., squares) as special cases or subsets of other classes (e.g., rectangles or rhombuses). At the same time, they should develop the more precise language needed to communicate ideas such as that all squares are rectangles but not all rectangles are squares. Students in the middle grades should also investigate what properties of certain shapes are necessary and adequate to define the class; they might explore, for example, the following question: Among the many characteristics of rhombuses, including congruent sides, opposite sides parallel, opposite angles congruent, diagonals that bisect each other, and perpendicular diagonals, which characteristics can be used to define rhombuses and to differentiate them from all other quadrilaterals? In a similar manner, other concepts introduced informally in the lower grades, including *congruence* and *similarity*, should be established more precisely and quantitatively during the middle grades, and special geometric relationships, including the Pythagorean relationship and formulas for determining the perimeter, area, and volume of various shapes, should be developed and applied. All these explorations should be carried out with the aid of hands-on materials and Dynamic Geometry software, and all should be conducted in an environment in which students are expected and encouraged to make and test conjectures and develop convincing arguments, based on both inductive and deductive reasoning, to justify their conclusions.

By the time students reach high school, they should be able to extend and apply the geometric knowledge developed earlier to establish or refute conjectures, deduce new knowledge from previously established facts, and solve geometric problems. They should be helped to extend the knowledge gained from specific problems or cases to more-general classes of objects and thus to establish the validity of geometric conjectures, prove theorems, and critique arguments proposed by others. As they do so, students should organize their knowledge systematically in order to understand the role of definitions and axioms and to appreciate the connectedness of logical chains, recognizing, for example, that if a result is proved true for an arbitrary parallelogram, then it automatically applies to all rectangles and rhombuses.

Specifying locations and describing spatial relationships

The importance of location and spatial relationships becomes apparent when we try to answer questions such as Where is it? (location), How far is it? (distance), Which way is it? (direction), and How is it oriented? (position). Typically, the first answers that children give to questions such as these are in relation to other objects: on the chair, next to the book, under the bed. In the primary grades, teachers help students develop a sense of location and spatial relationship by developing those early ideas. Using physical objects, often to illustrate stories, or physically acting out a relationship, children learn the meaning of such concepts as above, below, in front, behind, between, to the left, to the right, next to, and other relative positions. In time they add concepts of distance and direction, such as three steps forward, and they learn to combine such descriptions to lay out routes (e.g., walk to the door, turn left, go to the end of the corridor). Students begin to represent such physical notions of location, distance, and space both as verbal instructions and as diagrams or maps, and they learn to follow verbal directives and to read maps as means to locating a hidden object or reaching a desired destination. As their skills in representing locations increase, students should add more quantitative details by, for instance, pacing off or measuring distances to better communicate "how far" or adding a simple coordinate system to define a location more precisely.

In grades 3–5, students' understanding of location, direction, and distance are applied to increasingly more complex situations. They become more precise in their measurements and begin to examine situations to determine whether there is more than one route between two points or if there is a shortest distance between them. During these years, students should come to recognize that some positional representations are relative (e.g., *left* or *right*) or subjective (e.g., *near* or *far*), whereas others are fixed (e.g., *north* or *west*) and unambiguous (e.g., *between*) and that directions are not always interchangeable (e.g., two blocks north, then three blocks west does not take you to the same destination as two blocks west, then three blocks north, but three blocks west, then two blocks north does have the same end point as two blocks north, then three blocks west). They also should become more attentive to orientation; they might determine the direction that an object faces, whether it has been reflected or rotated from its initial position, or the distance and direction that it has been moved, all of which are closely related to ideas of transformations discussed later. It is particularly convenient and appropriate for students to explore the concepts of location and position by using grids together with graphical representations, physical models, and computer programs; as they continue their explorations on a grid, students also should learn to specify ordered pairs of numbers to represent coordinates and to use coordinates in locating points, describing paths, and determining distances along grid lines.

In the middle grades, the ideas established in elementary school should continue to be developed, and in addition, geometric ideas of location and distance can be linked to developing algebraic concepts as students apply coordinate geometry to the study of shapes and relationships. For example, the study of linear functions in algebra is related to the determination of the slopes of the line segments that form the sides of polygons, and these values in turn are used to determine relationships such as parallelism or perpendicularity of sides, which are used to analyze and classify the polygons; the Pythagorean relationship is applied to the coordinate plane to establish a method of determining the distance between points or the lengths of segments; and coordinates can be used to locate the midpoints of segments.

High school students should extend the geometric concepts of Cartesian coordinates used in lower grades to other coordinate systems, including polar, spherical, or navigational systems, and use them to analyze geometric situations. They also should develop facility in translating between different coordinate representations and should understand that each representation offers certain advantages in specific situations. During the secondary school years, students also learn to apply trigonometric relationships to solve problems involving location, distance, direction, and position, and analytical methods continue to be used, further strengthening the connection between algebra and geometry.

Applying transformations and symmetry

Among the early geometric discoveries that children make is that shapes can be moved without being changed: a triangle is still the same triangle even if it is flipped over or slid across the table, and a puzzle piece may need to be turned in order for it to fit into the desired space. Such intuitions are the starting point for studying transformations when children enter school. This important aspect of spatial learning in the primary school years engages students in exploring the motions of slides, flips, and turns, which leads to the discovery that such motions alter an object's location or orientation but not its size or shape. Primary-grades teachers should guide children to look for, describe, and explore symmetric shapes, which they can do informally by folding paper, tracing, creating designs with tiles, and investigating reflections in mirrors. Explorations that children enjoy in the primary grades, such as folding the net of a rectangular prism to make a "jacket" for a block, also involve relationships between two- and three-dimensional shapes.

As children move into the upper elementary grades, the more informal notions of slides, flips, and turns are treated with greater precision as translations, reflections, and rotations, and attention is directed to what parameters must be specified in order to describe those transformations (e.g., slide [translate] the square ten centimeters to the right; flip [reflect] the triangle over its hypotenuse; turn [rotate] the drawing a quarter-turn clockwise). Students also learn that transformations can be used to demonstrate that two shapes are congruent if one can be moved so that it exactly coincides with the other. They should then be helped to extend that notion by being challenged to visualize and mentally manipulate shapes, describing mathematically a series of motions that can be used to demonstrate congruence or predicting the result of certain transformations before actually performing them with physical objects or symbolic representations. Such growing precision extends as well to symmetry as students learn to specify all the reflection lines or the center and the degrees of rotation in a symmetric figure or design.

In grades 6–8, transformation geometry can be a powerful tool for exploring spatial and geometric ideas. Not only are rigid transformations used to deepen students' understanding of such concepts as congruence, symmetry, and the properties of polygons, but dilations and the notions of scaling and similarity, which are closely linked to proportional reasoning, are introduced in the middle grades. Additionally, in their study of transformations, middle-grades students should, for instance, generalize the result of two successive reflections over parallel lines and compare that outcome to the result of two successive reflections over intersecting lines. They should also begin to quantify and formalize aspects of transformations, establishing, for example, that in a reflection each point on the original object is the same distance from the mirror line as the corresponding point on the image. Physical manipulatives, such as mirrors or other reflective devices, and Dynamic Geometry software are especially useful in conducting such investigations.

In high school, the study of transformations can be further enriched by the use of function notation, coordinates, vectors, and matrices to describe and investigate transformations, including both isometries and dilations. Students should develop certain basic "tools" such as determining a matrix representation for accomplishing a reflection over the line $y = x$ or other common transformations, and they should relate the composition of transformations to matrix multiplication and apply those concepts to the solution of problems.

Using visualization, spatial reasoning, and geometric modeling

The ability to create mental images of two- and three-dimensional objects, to visualize how objects appear from different perspectives, to formulate representations of how objects are positioned relative to other objects, to relate two-dimensional renderings to the three-dimensional objects that they represent, to predict how appearances will vary as the result of one or more transformations, and to create spatial representations to model various mathematical situations are among the most important outcomes of the study of geometry.

Young children begin to develop their spatial visualization by initially manipulating physical objects and later extending their manipulations to mental images. Teachers in the primary grades may help children develop spatial memory and spatial visualization by asking them to recall and describe hidden objects or by having them describe how an object would look if viewed from a different side. They may ask children to imagine, and later explore and verify, what will happen when a given shape is cut in two in a certain way or to predict and demonstrate what other shapes could result if that same object were cut in a different manner. Children should also experiment with different shapes and formulate descriptions of them, perhaps by creating a shape from tangrams and taking turns to describe what each one sees in the figure. Students should also learn to read and draw simple maps and to give and follow directions—for example, giving a classmate verbal instructions for going from the classroom to the cafeteria. Opportunities abound to develop spatial visualization in connection with other topics and subjects by, for instance, demonstrating that even numbers can always—whereas odd numbers can never—be arranged in two equal rows or highlighting spatial concepts during art or physical education lessons. Children should have ample opportunity to discover that spatial reasoning and geometric modeling can contribute to understanding and solving a wide variety of problems that involve number, data, and measurement and that have numerous applications.

As students move through grades 3–5, they become more adept at reasoning about spatial properties and relationships among shapes; they might develop strategies to calculate the area of a garden plot by subdividing it into component rectangles or relate the area of a trapezoid or a parallelogram to the area of the rectangle that is formed by cutting and reassembling the original quadrilateral. Relating three-dimensional shapes to their two-dimensional representations becomes an important topic in these grades as students discover how to build three-dimensional objects from two-dimensional representations, and vice-versa; construct and fold nets of solids; examine diagrams of nets to predict which ones can or cannot be folded to form a certain prism; or mentally manipulate a shape to produce an accurate picture of hidden parts. Applying geometric reasoning and modeling to solve problems in all areas of mathematics, as well as in other contexts, should continue to be a principal focus of the curriculum.

The skills of spatial visualization and geometric reasoning that emerge in the lower grades should become more refined and sophisticated in grades 6–8 as students solve problems involving distance, area, volume, surface area, angle measure, and other quantifiable properties. Students should be guided to develop, understand, and apply important formulas for calculating the length, area, or volume of selected shapes. As they explore relationships using physical models and appropriate technology, students should begin to establish and give arguments to support important generalizations; they might, for example, demonstrate why, when the side of a cube is tripled, the surface area of the enlarged cube is nine times the surface area of the original whereas the volume of the enlargement is twenty-seven times that of the original. Geometric models for algebraic and numerical relationships help students integrate important concepts from all strands of the mathematics curriculum; manipulatives and computer programs that connect geometric, algebraic, and numerical concepts contribute to students' developing mathematical maturity and enable them to solve more-complex problems both within mathematics and in other subjects.

As students progress through high school, their visualization skills should extend from representations on the familiar two- and three-dimensional rectangular coordinate systems to analogous representations on a spherical surface or in a spherical space; investigations that connected two-dimensional representations of polyhedra with three-dimensional representations of them later evolve into challenges of projecting a spherical surface onto a plane and producing a two-dimensional map of a three-dimensional surface. Producing perspective drawings, visualizing the resultant cross section when a plane slices a solid object, predicting the three-dimensional shape that results when a plane figure is swept 360 degrees about an axis, and navigating in a spherical frame of reference are examples of spatial ideas that should evolve as students progress through school. In high school, too, geometric representations can be of great benefit when studying topics involving algebra, measurement, number, and data, and the application of geometric ideas to the solution of problems across mathematics and in other disciplines is one of the major goals of the curriculum.

Developing a Geometry Curriculum

A curriculum that fosters the development of geometric thinking envisioned in *Principles and Standards* must be coherent, developmental, focused, and well articulated, not simply a collection of lessons or activities. Geometric ideas should be introduced in the earliest years of schooling and then must deepen and expand as students progress through the grades. As they move through school, children should receive instruction that links to, and builds on, the foundation of earlier years; they must continually be challenged to apply increasingly more sophisticated spatial thinking to solve problems in all areas of mathematics as well as in other school, home, and life situations.

These Navigations books do not attempt to describe a complete geometry curriculum. Rather, the four *Navigating through Geometry* books illustrate how selected "big ideas" of geometry develop across the prekindergarten–grade 12 curriculum. Many of the concepts presented in these geometry books will be encountered again in other contexts related to the Algebra, Number, Measurement, and Data Standards; in the Navigations books, as in the classroom, the concepts described under the Geometry Standard reinforce and enhance students' understanding of the other strands.

Geometry is essential to the vision of mathematics education set forth in *Principles and Standards for School Mathematics* because the methods and ideas of geometry are indispensable components of mathematical literacy. The *Navigating through Geometry* books are offered as guides to help educators set a course for successful implementation of the very important Geometry Standard.

GRADES 3–5

NAVIGATING *through* GEOMETRY

Chapter 1
Shapes

Principles and Standards for School Mathematics (NCTM 2000) explains that in grades 3–5, students are beginning to expand their reasoning powers and thus can carry out more-complex explorations of the shapes they were introduced to in prior grades. They should develop more-precise ways of describing shapes, by focusing on identifying and describing their properties. For example, instead of describing a square as a four-sided figure whose sides all have the same length, they should be exploring the relationship among the angles and discovering the parallel and perpendicular relationships among the sides. Students should also be discovering the relationships among different shapes—how squares and rectangles compare, for instance. *Principles and Standards* points out that precision and clarity must be emphasized in using geometric terms in the mathematics classroom. When discussing shapes, students should be increasing their mathematical vocabulary by hearing terms used repeatedly in context. New terms, including *parallel*, *perpendicular*, *face*, *edge*, *vertex*, and *trapezoid*, must be used often and with precision so that they become part of the common vocabulary students use to communicate geometric ideas.

In prekindergarten–grade 2, students should have explored the properties of shapes through interacting with manipulatives in many free-play settings. For example, they may have handled cylinders and cubes and discovered that one rolls and the other does not. In grades 3–5 students still need many manipulative experiences to sort and classify, although their explorations should be more directed. Both examples and nonexamples of two- and three-dimensional objects should be studied in order to begin to classify shapes. Teachers should pay close attention to the vocabulary used, encouraging correct mathematical language so that geometric misconceptions are avoided. Students often call a three-dimensional shape by its two-dimensional name, for instance, using *triangle* to refer to a pattern block that is actually a triangular prism. In the spirit of exploration, teachers should always challenge students to look for patterns and generalizations with such questions as How do you know that this shape is not a rectangle? and What changes would be needed to make it a rectangle? By fifth grade, students should be able to compare and contrast a variety of two- and three-dimensional shapes and list the attributes and defining characteristics of the shapes.

The activities in this chapter help develop the mathematics of shapes for students in grades 3–5, as presented in *Principles and Standards for School Mathematics.* The first activity, Build What I've Created, is offered as an introduction to a unit on shapes. It affords students the opportunity to use the vocabulary they have acquired and thus gives the teacher a means of preassessing their knowledge. The activity also takes students further as they analyze their designs by combining and dividing shapes, leading to important discoveries about the properties of, and relationships among, shapes.

The next two activities focus on in-depth investigations of three- and four-sided figures. In Thinking about Triangles, students discover different types of triangles and their various properties. Students investigate four-sided figures in Roping In Quadrilaterals. These activities are truly exploratory, so students should be allowed time to make discoveries about the properties of shapes rather than be told about them. Reasoning skills are brought to bear heavily in both activities. Students are repeatedly asked to compare and contrast shapes, to categorize them, and to revise their categories. Such exercises help students gain a deeper understanding of the big ideas about shapes that were introduced in prekindergarten–grade 2.

Building Solids and Searching for the Perfect Solids focus on three-dimensional shapes. As *Principles and Standards* points out, geometry requires thinking and doing. It is important for students at this age to develop reasoning skills while building, tracing, measuring, and constructing three-dimensional models. The final activity, in particular, focuses on reasoning abilities, as students make a list of properties using examples and nonexamples of Platonic solids.

The five activities in chapter 1 will help students develop, understand, and use new vocabulary; explore properties and relationships of two- and three-dimensional shapes; and expand their reasoning skills by making and testing conjectures and justifying conclusions. Having completed the activities, they will have a much stronger understanding of shapes and be able to communicate mathematically with greater clarity, and they will have increased their mathematical-reasoning powers.

The first activity can serve as an introduction to the study of position and geometric shape. As you listen to students giving directions and observe the re-creation of the designs, you can gain insights into their knowledge of shapes, any misconceptions they might have, and the level of precision and clarity of their language. In the activity, students also combine and divide shapes into new shapes, which actually involves them in exploring the relationships between shapes, an important part of their growth in understanding geometry in grades 3–5. Once the activity has been completed, you have an excellent starting point for your geometry investigations, whether you wish to introduce the trapezoid in grade 3 or study the relationships between regular hexagons and equilateral triangles in grade 5.

Build What I've Created

Grades 3–5

Goals

- Construct a geometric design from oral directions
- Use precise geometric vocabulary in giving directions
- Recognize geometric shapes and patterns in quilt designs

Investigate, describe, and reason about the results of subdividing, combining, and transforming shapes

Prior Knowledge

Students should have a knowledge of such basic shapes as the triangle and the square and be able to use the attributes of these shapes to describe specific examples of them to a partner. It is not necessary that they know the formal terms for trapezoid and hexagon. They can learn them as the lesson progresses. It will be helpful if students have had previous experiences with pattern blocks.

Materials and Equipment

- Two identical sets of pattern blocks, each containing from six to twelve blocks and a variety of shapes for each pair of students. (A template for pattern blocks is available on the CD-ROM that accompanies this book.)
- A six-inch-square outline for each student on which to create designs (A template for this outline, "Quilt-Patch Work Space," is available on the CD-ROM.)

Learning Environment

Students work in pairs. If the partners are facing each other, place a divider between them so that the builder cannot see the pattern being described. Otherwise, they can sit back to back. The activity is most successful when students are paired homogeneously by ability in both verbal and spatial skills. This activity can be easily modified by altering the number of blocks used and by varying the types of questions the students are allowed to ask. Most students should be encouraged not to use color to identify a block but rather to use its name or talk about its characteristics. For advanced students, the re-creator of the design might be allowed to request only that the directions be repeated. Other students might be permitted to ask only questions that can be answered by yes or no. Students who need more help could have no restrictions on dialogue between the direction giver and the re-creator of the design.

Important Geometric Terms

Shapes: *triangle, square, hexagon, trapezoid, rhombus*

Position: *above, below, on top of, left, right, parallel, perpendicular, horizontal, vertical,* and *adjacent*

To explore cultural connections, students could read about the history of quilts in the United States in a classroom reading center on this topic. The Web site www.womenfolk.com is useful. Go to Grandmothers' Quilts for a history of quilts and photographs of different kinds of quilts.

Using the Pattern Patch applet on the CD-ROM, students can design a quilt patch and explore characteristics of shapes.

Activity

Engage

Quilting activities can be used to make many connections to literature. You could begin this activity by reading *Sweet Clara and the Freedom Quilt*, by Deborah Hopkinson (1993), or *Sam Johnson and the Blue Ribbon Quilt*, by Lisa Ernst (1983). Besides portraying positive gender-role models for students, both books have excellent connections to social studies. Discuss the quilt patterns in the book, encouraging the students to determine how they are alike and how they are different, and note the connection to mathematics and art.

Explore

On the "Quilt-Patch Work Space," one student designs a "quilt patch" from pattern blocks (fig. 1.1) without the partner's seeing it. Then the creator of the design gives directions to the partner so that the partner can re-create the design without looking at it. The re-creator then explains to the designer what the design looks like. Finally, the quilt designs are compared visually, and the students together record answers to the following questions:

- What words or phrases helped you re-create the design?
- What words or phrases confused you? Why?
- Can you think of better ways to explain the directions for making the design?

The activity is then repeated with partners switching roles. Follow the activity with a classroom discussion that develops the geometric vocabulary used by the students.

Have the students examine the designs to see if they can rearrange the pattern blocks to find new shapes with five, six, or seven sides. They should record which blocks make up the new figure, trace around the new shapes, and verify the number of edges by counting the sides in the diagram. They can also discuss the number of angles in their diagrams and compare that number with the number of sides.

Fig. **1.1.**

A "quilt patch" composed of pattern blocks

Assess

This activity can be used for preassessment at the beginning of a unit on geometry. By observation, teachers can determine students' familiarity with the geometry terms that are important in this activity.

To assess students' performance, ask them to design a quilt patch and then write directions for another student to re-create the design. Their success is determined by the quality of another student's re-creation of the design. This approach encourages self-assessment as students modify their directions on the basis of the outcome.

This assessment activity can be set up as a classroom center. The designs can be used as a decorative room or door border showing the connection of mathematics to art.

Extend

Show students prints of paintings, for example, *Harmonie Tranquille*, by Wassily Kandinsky, or *Three Musicians*, by Pablo Picasso, and ask them to identify all the different shapes found in the paintings. Working with the art teacher, you can introduce cubism to students and encourage them to create original artwork that includes a variety of shapes in this genre. You can find copies of these paintings at www.artmodes.com.

Where to Go Next in Instruction?

With this activity as an introduction, students should be prepared to begin a unit on discovering the properties of shapes and angles. Be flexible in grouping students on the basis of informal observations of their performance in this activity. Pattern blocks are excellent manipulatives for students to use to explore combining and transforming shapes. Spark students' discovery with the following questions:

- How many ways can you make a quadrilateral by putting different blocks together? Show some examples.

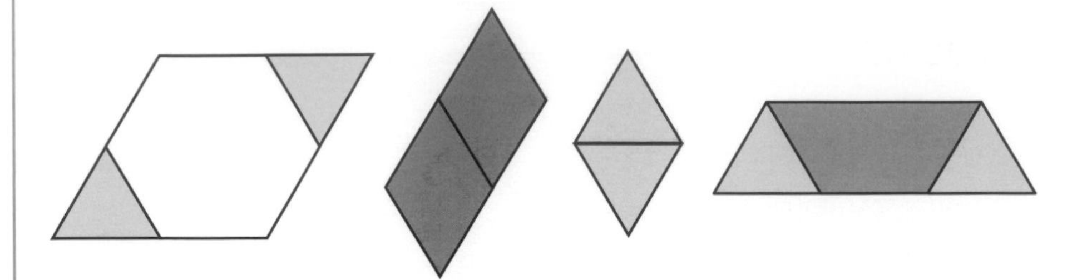

- What are the names for the quadrilaterals you made?
- Which ones are the same size and shape (congruent)? How can you tell?
- Can you make a five-sided figure by combining shapes? A six-sided figure? A seven-sided figure?
- What patterns do you notice?

Students may notice that in combining shapes, they lose one side of each shape. So if they want to make a pentagon (five sides), they can use a triangle (three sides) and a square (four sides). In combining the triangle and square, one side of each figure is "lost" when it becomes part of the interior of the pentagon.

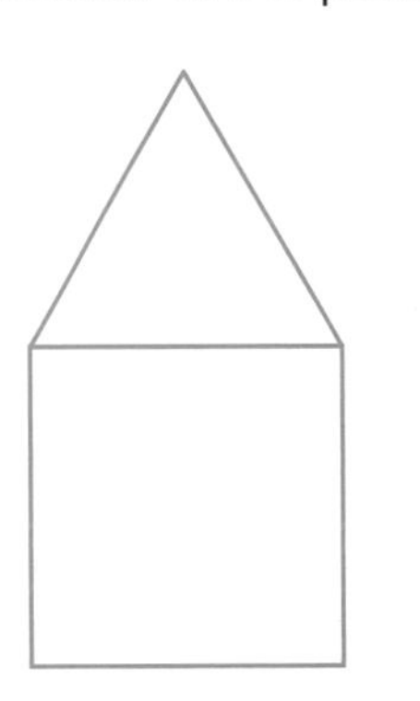

Pattern blocks are also excellent manipulatives for exploring angle measures. For example, three green triangles can be put together at one of their vertices to demonstrate that the angle formed is a straight angle (measure 180°); thus, each of the three angles measures 60 degrees. The measure of each angle of a hexagon is made of two angles of the green triangle, so each interior angle of a hexagon measures 120 degrees.

Teachers should be aware that pattern blocks are actually three-dimensional shapes; however, they are commonly referred to by the name for the two-dimensional shape of the two large faces. For example, the triangle is really a triangular prism. It is not necessary to elaborate on this distinction at this time, but be aware of it in case a student recognizes it.

Quilt designs can also be used to explore symmetry and transformations. Patchwork Symmetry in chapter 3 offers further explorations and makes a nice segue from the study of shapes into the study of transformations, using quilting as the link.

Thinking about Triangles

Grades 3–4

Goals

- Use geoboards to investigate properties of triangles
- Transcribe geoboard designs onto geodot paper
- Make and test conjectures about properties of triangles
- Draw conclusions on the basis of experimentation with a variety of examples

Identify, compare, and analyze attributes of [two-dimensional] … shapes and develop vocabulary to describe the attributes

Make and test conjectures about geometric properties and relationships and develop logical arguments to justify conclusions

Prior Knowledge

Students should know the definition of *polygon* (see "Important Geometric Terms") and be able to recognize and draw triangles, to recognize right angles, and to compare the sizes of angles (e.g., discover that one is greater than another). If students are using geoboards for the first time, a period of free exploration should be allowed so they can become familiar with this tool.

Materials and Equipment

- A five-pin-by-five-pin geoboard with one rubber band for each student
- A copy of the blackline master "Geodot Paper for Geoboards" for each student
- Paper, pencils, and crayons for recording and writing journal reflections

Geoboards allow students to explore their conjectures quickly and easily; however, "Geodot Paper for Geoboards" can be used as an alternative if necessary.

p. 100

Students can also use the Geoboard applet on the CD-ROM.

Learning Environment

For this activity, students could work in small groups to encourage mathematical discourse while they are making their shapes and testing their conjectures. Depending on the group of students and the question you are exploring, however, you may wish to have the students work alone at first to make a conjecture and begin testing it. After they have done some exploration, they can come together to discuss their findings in groups of two to four and then draw and justify conclusions together. In any event, it is best if the students have their own geoboards to be able to construct their own ideas about the properties of shapes.

Important Geometric Terms

Right angle, *acute angle*, *obtuse angle*, *equilateral triangle*, *isosceles triangle*, *scalene triangle*

Polygon: A closed plane figure formed by three or more line segments called *sides*. Each side intersects exactly two other sides, but only at

---

This activity has been adapted from Walker, Reak, and Stewart (1995a), *Twenty Thinking Questions for Geoboards*.

their end points. Sides that have common end points are not part of the same line.

Activity

Engage

Ask the students to search the classroom to find triangular objects and list them. Ask such questions as the following:

- What do these shapes have in common?
- Where do you think the word *triangle* comes from?
- What other words do you know that start with *tri?* (tricycle, tripod)
- How are they similar to the word *triangle?* How are they different?

Explore

The explorations are a series of questions for students to conjecture about, explore using their geoboards, and discuss in small groups before finally coming to a class consensus through guided discourse. Each question is expected to take a full one-hour class period. Encourage students' discussion in their groups and pose questions rather than give answers when observing the groups in action. For example, rather than tell a group that their figure is not a polygon, ask them why they think it is a polygon and then together discuss the definition of a polygon. You might even show them examples of polygons and examples of nonpolygons and ask them in which category their figure belongs.

Question 1: Is it possible to make a three-sided polygon that is not a triangle?

The students first make a conjecture (yes or no) and record and tally all the students' conjectures. They then use their geoboards to explore possibilities and come to a consensus in their groups. Groups of up to four work well for this task.

If the students construct shapes like those in figure 1.2, refer them to the definition of a polygon and ask, "Is your figure a closed figure with no lines crossing?" Conversely, if they make a triangle and do not identify it as such, refer them to the origin of the word triangle (three angles). Ask, "Does your shape have three angles?"

As a closing discussion, make a class list of all the properties of triangles. The students should mention three angles and three sides. They

Fig. **1.2.**
Examples of nonpolygons

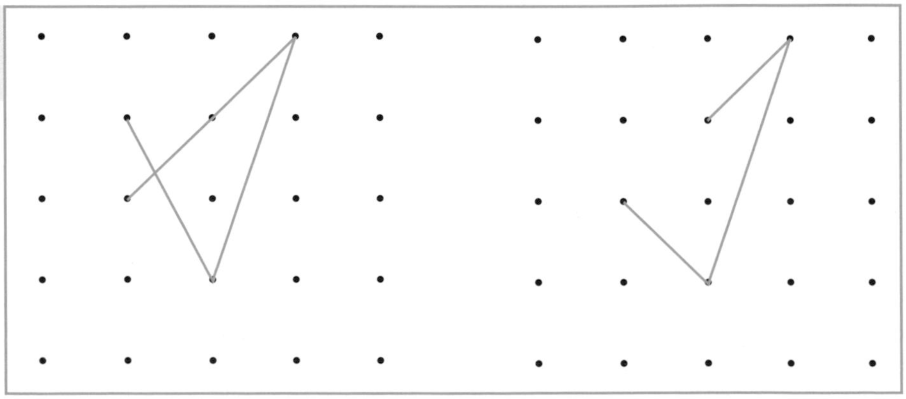

may also state that a triangle is a polygon. Alternatively, have the students write a journal reflection for assessment.

Question 2: Is it possible for a triangle to have two right angles?

Again the students make individual conjectures and tally the class's responses. They then test their conjectures on the geoboard and come to a consensus in small groups. Follow up with a whole-class discussion. If the students construct a figure that has two right angles but is not a triangle (see fig. 1.3), ask them if their figure has all the properties of a triangle.

To test for right angles, the students can use the corner of an index card or take any rectangular piece of paper and fold it in half and then in half again. The meeting of the two folds forms a right angle.

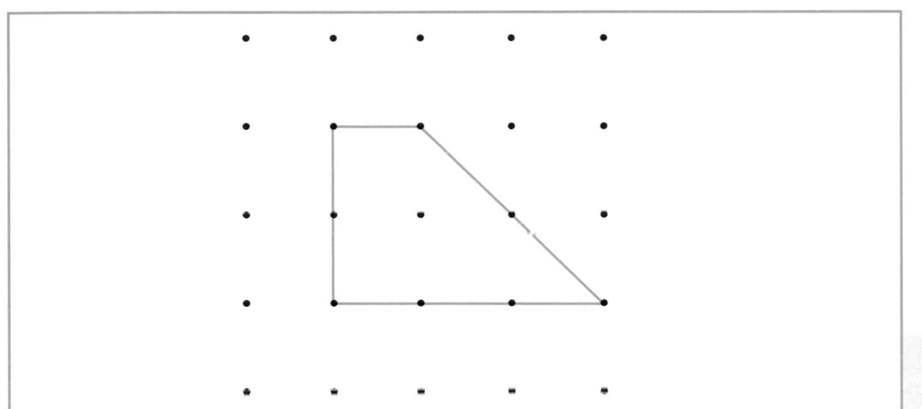

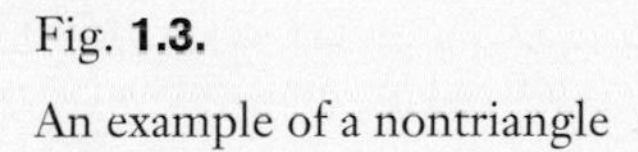

Fig. **1.3.**
An example of a nontriangle

Question 3: How many different right triangles can be made on the geoboard?

Have the students make and tally conjectures, as before. In introducing this question, discuss what *different* means. For this task, if a triangle can be flipped or turned to match one that has already been made, then it is not "different."

This is an intuitive introduction to the concept of congruence (having the same size and shape). To test for congruence, the students may need to cut out triangles or use two transparencies to overlay triangles.

The students should record all the triangles on geodot paper and explain how they know they are right triangles. If the students find only a few right triangles, encourage them to search for more. Ask them, "Are you sure you found all the right triangles? What strategy did you use to make sure you have found them all?"

Teachers are advised to attempt this task before assigning it to their students so that they can devise a strategy and experience the discovery and thinking processes that their students will use. The fourteen right triangles that can be constructed on the five-pin-by-five-pin geoboard are shown in figure 1.4. If the students are having difficulty coming up with a strategy to find them all, model your approach. For example, "I started with a right triangle that had a base of one unit and a height of one unit. I kept the base the same and tried to find how many more right triangles I could make. Then I made one with a base of two units.…"

Question 4: How many different types of triangles can you find?

Have the students make and tally conjectures, as before. Thus far, this activity has focused on right triangles. Model examples of one or two other types (e.g., an isosceles triangle or a scalene triangle). Encourage the students to think about the sizes of the angles and the

Fig. **1.4.**

The fourteen right triangles that can be made on a five-pin-by-five-pin geoboard

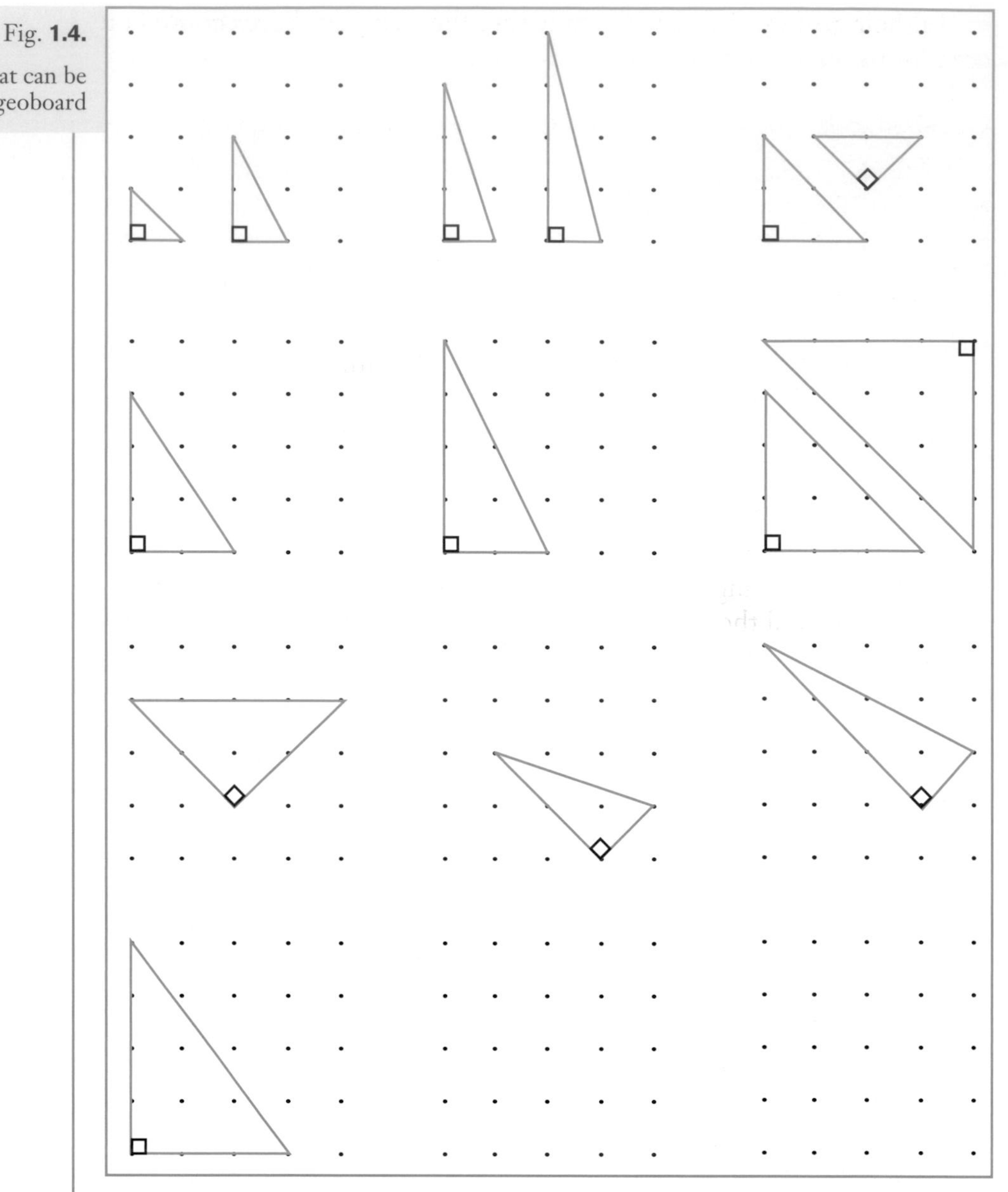

lengths of the sides of other triangles they can create. Have them record their triangles on geodot paper.

It is not possible to make an equilateral triangle on the geoboard. Some students may claim that the triangles in figure 1.5 are equilateral because the lengths of the sides look equal or because the band on the triangle on the right connects two pegs with one peg in between. Encourage them to measure the sides to prove that the triangles are not equilateral. Discuss the properties of equilateral triangles, and give examples.

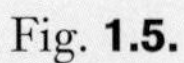

Fig. **1.5.**

Examples of students' unsuccessful attempts to construct equilateral triangles on the geoboard

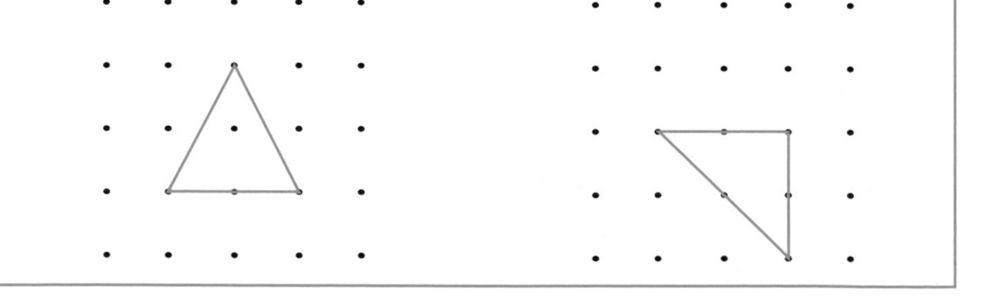

In their groups, the students can share the triangles they recorded on geodot paper, put into piles those that are the same, and then label each pile with a defining characteristic. Restrictions can be placed on the number of piles (e.g., there must be at least three piles) and the number of triangles within a pile (each pile must have at least three triangles in it). The students could create posters with triangles displayed by categories and make presentations to the class explaining the strategies they used to determine the classifications. They should be prepared to justify the labels and defend the placement of the triangles. Follow the presentations with a class discussion that introduces the terms *acute*, *obtuse*, *scalene*, and *isosceles*. Do not introduce these terms until after the students have made their presentations. It is very important that the students think up their own categories to describe the triangles. The actual geometric terms will evolve naturally from their classifications.

Assess

During each of the explorations, the teacher should circulate among the groups, observing and attending to students' conjectures and the reasoning behind them, students' discussions of congruence and of geometric terms, and their classifications for different triangles.

This activity lends itself nicely to written journal reflections from which teachers can determine what individual students have learned. The following can be used as prompts for reflections about each of the questions explored:

- Question 1: What have you learned about a triangle from this investigation?
- Question 2: If you could make a triangle that was as large as you wanted, would you be able to make one that had two right angles? Explain your thinking. (Examples of students' responses are shown in fig. 1.6.)
- Question 3: Write everything you know that is true about all right triangles.
- Question 4: Write in your own words the definitions for the new geometric terms we have found (*isosceles*, *scalene*, *acute*, and *obtuse*).
- Summary question: Finish each sentence with as many different answers as possible:

  All triangles have …
  Some triangles have …

Extend

Using straws of different lengths or The Geometer's Sketchpad (Jackiw 1991), students can explore the following questions:

- Can a triangle be made with segments measuring five, six, and seven units? Can more than one triangle be made? Why or why not?
- If you are given any three lengths, can you always make a triangle? Why or why not?
- Using several different sets of three lengths, try to make triangles. Can you make up a rule about the lengths of the sides of triangles?

In answering these questions, the students derive the theorem that the sum of the lengths of any two sides of a triangle must always be greater

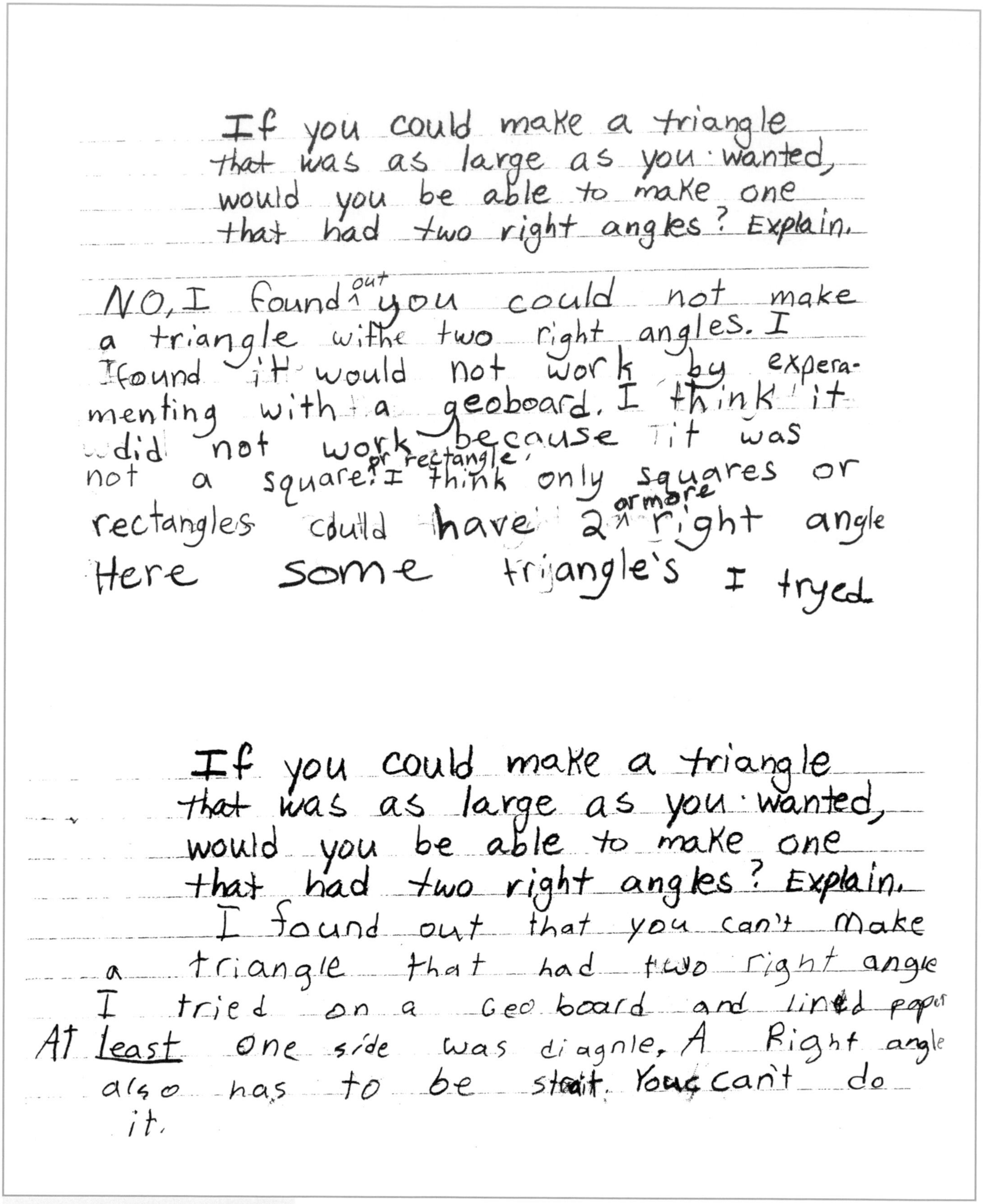

If you could make a triangle that was as large as you wanted, would you be able to make one that had two right angles? Explain.

NO, I found out you could not make a triangle withe two right angles. I Ifound it would not work by experamenting with a geoboard. I think it did not work because it was not a square or rectangle. I think only squares or rectangles could have 2 or more right angle Here some trijangle's I tryed

If you could make a triangle that was as large as you wanted, would you be able to make one that had two right angles? Explain.

I found out that you can't make a triangle that had two right angle I tried on a Geoboard and lined paper At least one side was diagnle. A Right angle also has to be strait. You can't do it.

Fig. 1.6.

Students' reflections on their attempts to create a triangle with two right angles than the length of the third side. Can a triangle be made with three sides of any length? A triangle can be made with sides of lengths five, six, and seven units, since $5 + 6 > 7$, $6 + 7 > 5$, and $5 + 7 > 6$. However, a triangle cannot be constructed with sides of lengths one, two, and three units, since $1 + 2$ is not greater than 3. The students are learning the importance of a counterexample to disprove an idea here. If they find one example that does not work, then the conjecture is invalid. Note that this investigation can also take place by putting together geostrips to test conjectures instead of using the computer.

Similar questions can be used to explore squares and other rectangles:

- How many different squares can you make on your geoboard?
- How many different rectangles can you make on your geoboard?
- How do you know that you have found them all?
- How do you know they are all different?
- How are a rectangle and square alike? How are they different?

In grades 4 and 5, students can proceed to investigate parallelograms and come up with the definition of a rectangle. To help them examine their definition, ask the following questions:

- Could a square be considered a special type of rectangle? In what way?
- Could a square be considered a special type of parallelogram?

A Venn diagram can then be developed to show the relationship among the three types of quadrilaterals (see fig. 1.7).

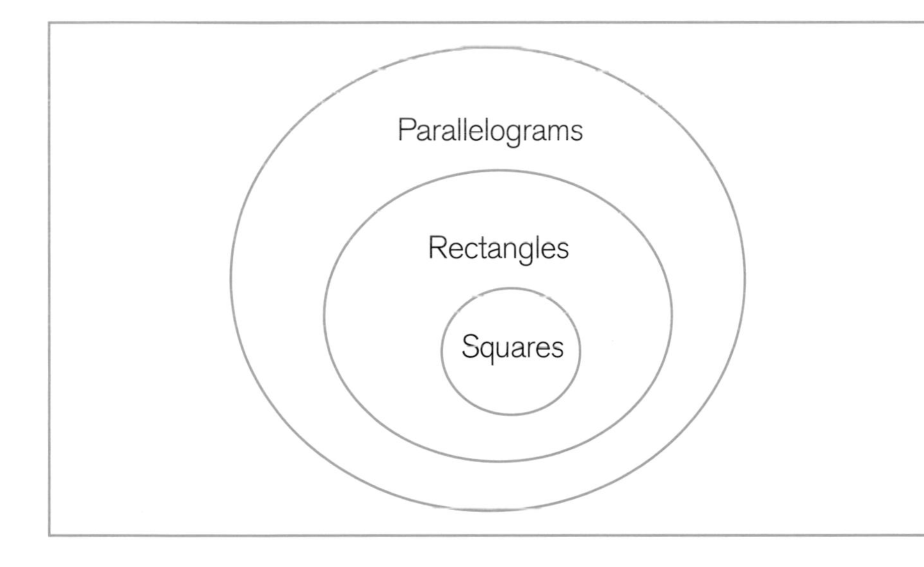

Fig. **1.7.**

A Venn diagram showing the relationship among three types of quadrilaterals

Roping In Quadrilaterals

Grades 4–5

Identify, compare, and analyze attributes of [two-dimensional] … shapes and develop vocabulary to describe the attributes

Classify [two-dimensional] … shapes according to their properties

Make and test conjectures about geometric properties and relationships and develop logical arguments to justify conclusions

pp. 101, 102, 103

The definitions of *trapezoid* differ in commercial textbooks. Most elementary-level textbooks define it as a quadrilateral with at least one pair of parallel sides. Others refer to a trapezoid as a quadrilateral with only one pair of parallel sides. The Navigations books use the latter definition.

Goals

- Sort quadrilaterals on the basis of specific attributes
- Use Venn diagrams to classify quadrilaterals
- Determine the common attributes of a set of quadrilaterals

Prior Knowledge

Students should be familiar with the characteristics of quadrilaterals, including parallelograms, trapezoids, rectangles, rhombuses, and squares. They do not need to have examined the relationships among these figures. The relationships will be developed during the activity.

Materials and Equipment

- Three pieces of yarn or string or three plastic hoops for each group of students
- A set of quadrilateral pieces for the teacher from the blackline master "Quadrilateral Pieces"
- A set of quadrilateral pieces from "Quadrilateral Pieces" for each group of students
- Ring labels for each group of students from the blackline master "Ring Labels"
- A set of mystery rings from the blackline master "Mystery Rings"
- Blank index cards and markers for each group of students
- Measuring tools, including rulers and index cards, for each group of students to test for right angles

Learning Environment

Students work in groups of three or four for the game "Questions, please!" and the "Explore" tasks.

Important Geometric Terms

Quadrilateral, parallelogram, rectangle, rhombus, square, trapezoid (see the note in the margin), *right angle, acute angle, obtuse angle, parallel lines, adjacent*

Kite: A convex quadrilateral with two pairs of adjacent sides that are equal in length

Activity

Engage

In order to familiarize students with the shapes used in this activity, play the guessing game "Questions, please!" Each group of students

---

Parts of this activity have been adapted from "Shape Up!" (Oberdorf and Taylor-Cox 1999).

should have a set of quadrilateral pieces spread out in front of them. Select one piece from your set of quadrilaterals, and hide it in your pocket or desk drawer. In order to determine which piece you have hidden, the groups in turn ask questions that can be answered by yes or no (e.g., Does the figure have four equal sides?) or make an actual guess by naming a particular quadrilateral piece. The team that holds up the specific piece on its turn is the winner.

This game is rich in opportunities to develop strategies and reasoning. Students must formulate good questions, that is, questions that will eliminate several pieces at once and narrow the possibilities. Note that the game can be won only by actually naming the piece. As the game proceeds, students must decide whether to risk naming a piece or ask a more general yes-or-no question. To take advantage of opportunities that arise to discuss these strategies with the students, ask questions such as the following:

- What is the best first question to ask? Why?
- Is there another question that is equally good? Why or why not?
- If you were left with a square, a rhombus, and a rectangle, what would be a good question to ask? (Does the shape have right angles? Does it have congruent sides?)

Explore

The students place the sixteen quadrilateral pieces in Venn diagrams they create from the three pieces of yarn or string or the three hoops and label the rings with the categories for each task on "Ring Labels." It is advisable for teachers to model task 1 (fig. 1.8), especially if the students are not familiar with Venn diagrams.

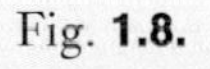
Fig. **1.8.**

The rings for task 1 with the quadrilaterals properly placed

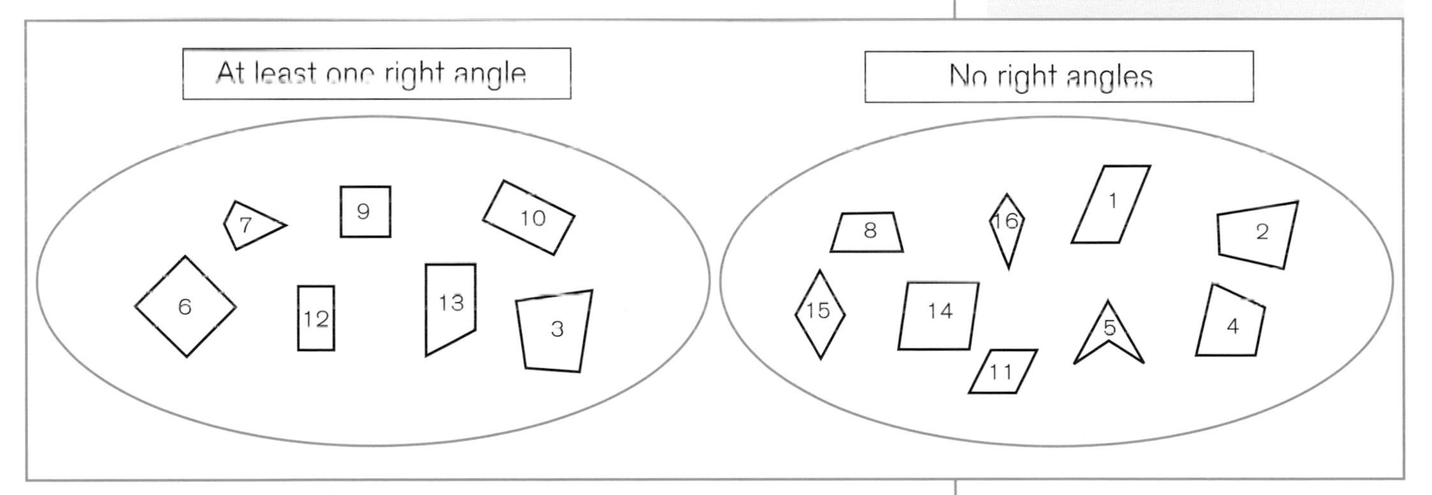

If the students have not previously worked with Venn diagrams, explain that the quadrilaterals that possess the characteristics for both rings should be placed in the intersection of the two rings. For example, if one ring were labeled "Right angles" and the other, "Congruent sides," a square would be placed as shown in figure 1.9.

Each group of students can paste the rings on a large piece of paper for viewing. A summary class discussion can generate good mathematical discourse about the defining characteristics of each quadrilateral and the relationships among the quadrilaterals. To facilitate this discussion, ask questions such as these:

A similar activity, Shape Sorter, is available on the CD-ROM.

Fig. **1.9.**

A square properly placed in the intersection of two rings

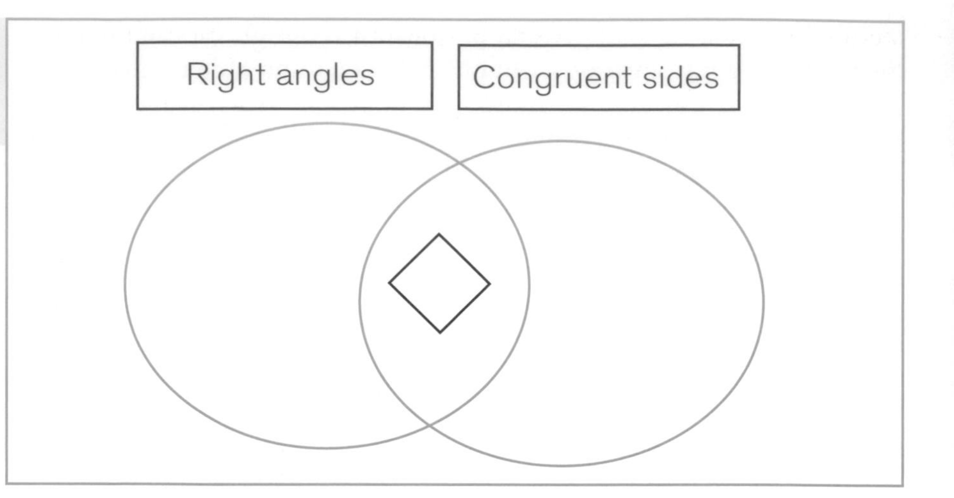

- Why did you place that shape in the intersection? What characteristics does it have?
- What do all the shapes in one ring have in common?
- How might the shapes in a ring be different?
- What different label would eliminate one or more of the shapes from a ring?
- What different label for one of the rings would allow you to include a new shape?

Assess

During the activity, circulate among the groups and ask students to defend their placement of different pieces. It is especially important to hear the students' reasoning for the placement of figures in the intersection of two or three rings. Ask the students why some of the figures should remain outside all the rings. Ask, "If we drew a giant circle around everything, including the shapes that are outside the rings, what might the label for this new ring be?" (quadrilaterals)

Use the questions listed in the "Explore" section to encourage the students to talk about the relationships among the figures. Talk about the placement of the figures in some of the displays. Have the children question and defend the placement. Journal reflections explaining the placement of quadrilaterals are useful for checking individuals' understanding.

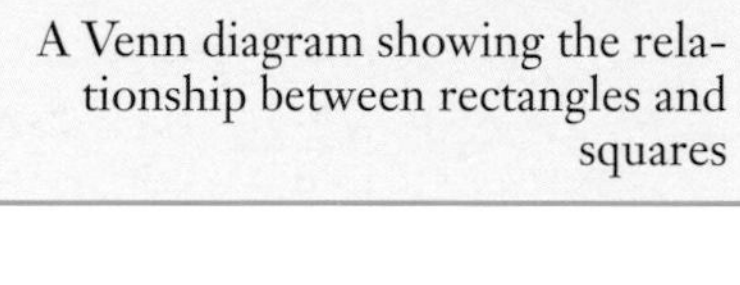

Fig. **1.10.**

A Venn diagram showing the relationship between rectangles and squares

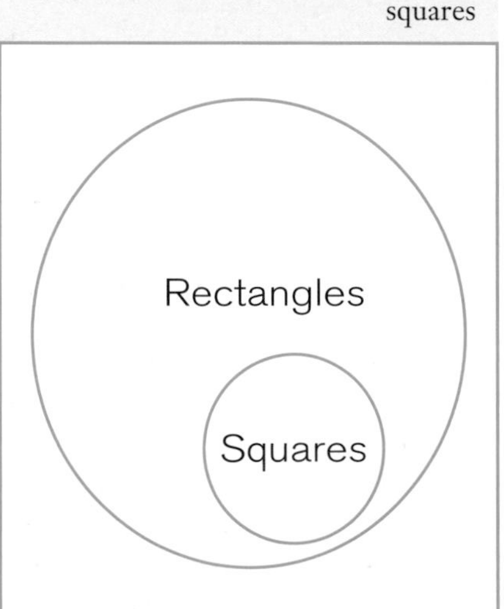

This activity affords an excellent opportunity to get students thinking about the relationship between a square and a rectangle. Younger children may have thought of them as two distinct shapes, but now they should discover the relationship between them. Ask the students to list the characteristics of a square and then a rectangle. Ask, "Is a square a rectangle?" Let the students think out loud about this question. Give them time to talk in small groups. Some students may argue that the two shapes are not related, since many rectangles do not have four equal sides. Refer them to the defining characteristics of a rectangle (four right angles, with opposite sides parallel); unequal sides are not mentioned in the definition. Then explore the question Is a rectangle a square? Students may benefit from the creation of a Venn diagram to show the relationship (fig. 1.10), which is a difficult one for students to understand fully. They need to see that every square is a rectangle because all squares satisfy the definition of a rectangle but that many rectangles are not squares.

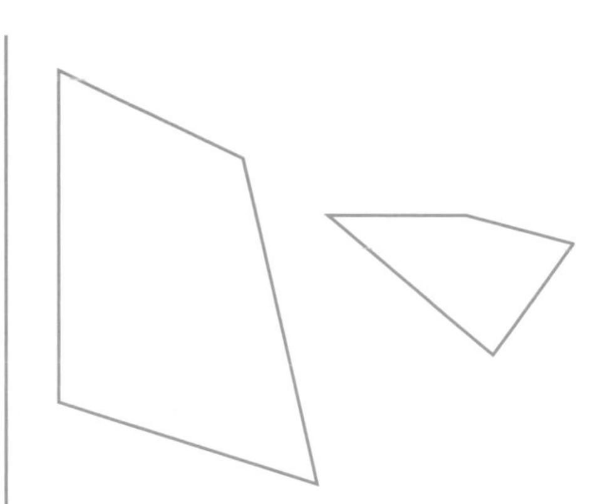

Extend

"Questions, please!" can be extended by giving the students some clues and having them come up with one or more figures that fit the clues. For example, ask them to draw a quadrilateral that has at least two sides equal in length and no sides parallel. Then have them make a different type of quadrilateral using the same clues. You can further extend the game to include all types of polygons.

Roping in Quadrilaterals can be extended by having students make up their own labels and then challenge a partner to use them to create quadrilateral rings. The blackline master "Mystery Rings" can be used to reverse the investigation: the students decide on appropriate labels for the rings, write them on index cards, and post them next to the correct rings. As a further extension, the students can make their own "mystery rings" for their classmates to label.

After much exploration with these activities, students can begin to look at the relationships among all types of quadrilaterals. One way is to classify quadrilaterals by the lengths of their sides:

- Having two pairs of opposite sides congruent (all parallelograms)
- Having two pairs of adjacent sides congruent (kites, squares, rhombuses)

The relationship among the rectangle, rhombus, and square should also be explored. A square is both a rectangle and a rhombus. A Venn diagram is a good visual aid to illustrate this relationship (see fig. 1.11.) Students at this age are just beginning to comprehend these classifications. They will benefit from thinking and talking about them, but they will need to continue to revisit the relationships before they solidify their understanding.

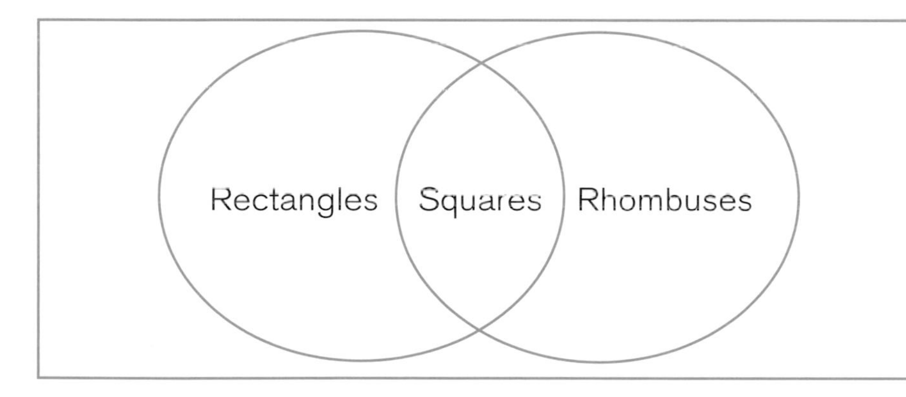

Fig. **1.11.**

A Venn diagram illustrating the relationship among rectangles, rhombuses, and squares

Building Solids

Grades 3–4

Goals

Make and test conjectures about geometric properties and relationships and develop logical arguments to justify conclusions

- Build a variety of solids from materials supplied
- Discover the relationship among the numbers of faces, edges, and vertices of solids
- Discover relationships between two- and three-dimensional figures

Prior Knowledge

Students should be familiar with properties (number of sides and angles) of many two-dimensional shapes, including triangles, squares, rectangles, pentagons, hexagons, and octagons. They also should be able to recognize congruent shapes.

Materials and Equipment

- A set of approximately sixteen "sticks"—for example, toothpicks or coffee stirrers—of each of four different lengths (two to eight inches)— for each pair of students
- A variety of "fasteners" (e.g., gumdrops, miniature marshmallows, clay)
- One copy for each pair of students of the blackline masters "Two- and Three-Dimensional Shapes" and "Counting Parts of Solids"
- For each pair of students, a set of three-dimensional wooden or plastic shapes—including one each of a cube, a square pyramid, a cylinder, a cone, and a sphere—and a variety of prisms such as a rectangular prism, a triangular prism, a pentagonal prism, a hexagonal prism, or an octagonal prism

If using marshmallows, allow them to harden by exposing them to the air for at least a day. Otherwise the figures are not rigid and become distorted.

pp. 104, 105

Important Geometric Terms

Cube, pyramid, prism, cylinder, sphere, cone

Polyhedra (singular, *polyhedron*): Solids whose faces are polygons

Polygon: A closed two-dimensional figure that is made up of line segments that intersect only at their end points

Faces: Polygonal regions that make up the surface of a solid

Edges: The line segments created by the intersection of two faces of a solid

Vertices (singular, *vertex*): The points of intersection of two or more edges

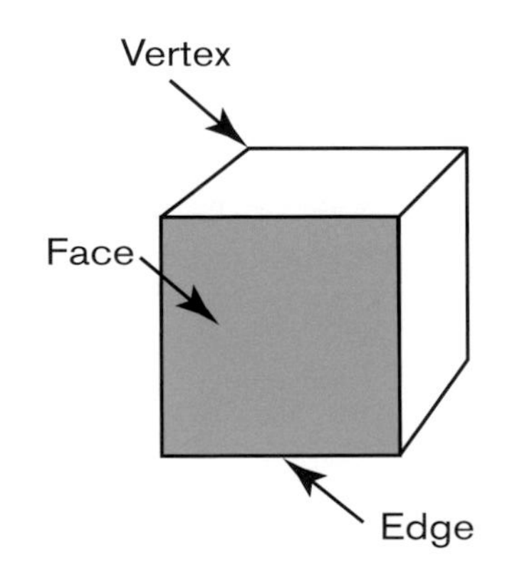

Learning Environment

Students work in pairs to construct models of solids.

This activity has been adapted from Battista and Clements (1998).

Activity

Engage

This activity will take more than one class period. In fact, plan on one class period for the "Engage" section alone. Set the stage with the introduction:

> Today we enter into a different dimension in mathematics—the third dimension. It is hardly new to you, since we live in the third dimension. But in class, we have been exploring shapes only in two dimensions. Two-dimensional shapes lie flat on a tabletop, and they can be drawn on a piece of paper. But three-dimensional shapes can stand up—they have height.

With the students' input, make a list of two-dimensional (2-D) shapes (triangles, squares, rectangles, circles, pentagons, etc.). Show the students the set of solid shapes. Ask pairs of students to explore the following questions: How are these three-dimensional (3-D) shapes like the ones we have written down? How are they different? Make a list of the similarities and the differences. Bring the class together to discuss their findings. (See an example of one pair's work in fig. 1.12.)

Next, distribute copies of "Two- and Three-Dimensional Shapes" and assign the following task to pairs of students:

> Select one 3-D shape and compare it to one or more 2-D shapes. Write down on the activity sheet all the things that are alike about the shapes and all the things that are different.

(One pair's list of likenesses and differences is shown in fig. 1.13.) Again summarize the exploration with a class discussion of the task.

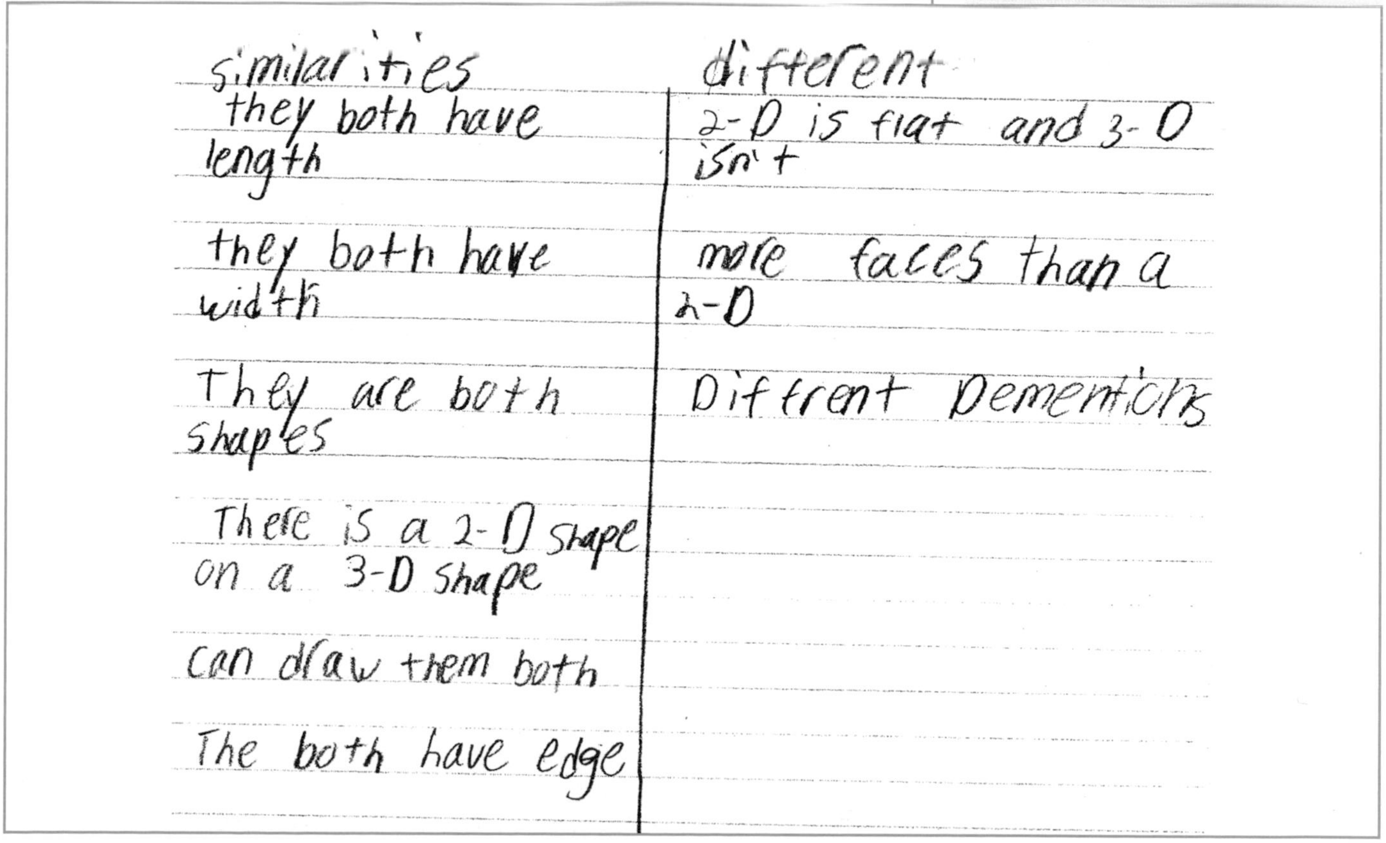

Fig. **1.12.**

A list of similarities and differences between two- and three-dimensional shapes

similarities	different
they both have length	2-D is flat and 3-D isn't
they both have width	more faces than a 2-D
They are both shapes	Diffrent Dementions
There is a 2-D shape on a 3-D shape	
can draw them both	
The both have edge	

Similar

1. A cube ir like a square because the cube seems like a layered square.
2. A cube is like a square because you see a square on each side of the cube.

Different

1. A cube is different from a square because a cube is bigger than asquare.
2. A cube is different from a square because the cube has more edges and facer, than a square

Fig. **1.13.**

A comparison of a 3-D shape and a 2-D shape

The figures that are impossible to build with the materials available have a circular shape (cone, sphere, and cylinder). The figures that are possible to build are polyhedra.

Explore

Pass out to each pair of students a set of "sticks" and a supply of "fasteners." Ask the students to create as many of the solids as they possibly can. You might suggest to them that they may find some solids that cannot be created.

Remind the students to try to build at least a cube, a rectangular prism, a triangular prism, and a square pyramid and to turn and flip their constructions to make sure they are truly different. You can refresh the students' understanding of congruence by referring to two-dimensional shapes if needed. After the models have been completed, have the students create a display of those they were able to build, putting them next to the appropriate three-dimensional block. Put all

the blocks they could not model together. Ask, "How are the figures you were able to build different from the ones you could not build?"

Next introduce the terms *faces*, *edges*, and *vertices* by referring to the terms used for two-dimensional shapes (e.g., sides are called *edges* and corners are called *vertices*). The 2-D shape becomes a face of the 3-D shape. A good way to remember the term *face* is that it makes up the sur*face* of the solid.

Have the students count the number of faces, edges, and vertices in each of the shapes they constructed. They can use the blackline master "Counting Parts of Solids" as a recording sheet.

p. 105

Children at this age may have difficulty counting the parts, especially the edges, because they have a difficult time finding an organized method for doing so. Rather than give them a strategy for counting, suggest that they devise a way that helps them avoid counting an edge or a corner twice. If some pairs of students continue to have trouble counting, have others share their successful strategies with the class. For example, they might label each vertex with a letter or color each edge they count.

Assess

Ask the students to write a response to the following in their journals:

> I am a square and have just been beamed into your very unusual world from my flat land of two dimensions. What a strange world you have here! There is a figure that sort of looks like me, but not really. It looks like the figure in the margin. Please tell me everything you know about this figure.

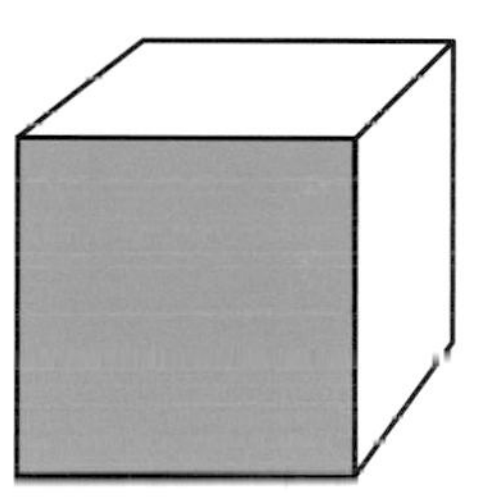

Similar scenarios can be created for cylinders and prisms.

As an alternative assessment, students can create pop-up books in which two-dimensional shapes pop off the page into three-dimensional shapes (a square becomes a cube, a rectangle becomes a rectangular prism, a circle becomes a cylinder, etc.). This can be a class project in which a story line is created and a "big book" is made. The students can read the book to children in grades 1 and 2.

How to Make Pop-Ups *(Irvine 1987)*, How to Make Super Pop-Ups *(Irvine 1991)*, *and "Making Connections: From Paper to Pop-Up Books" (Huse, Bluemel, and Taylor 1994) may be helpful in constructing these books. The latter is available on the CD-ROM that accompanies this book.*

Extend

On the accompanying CD-ROM or the NCTM Illuminations Web site (address: illuminations.nctm.org; click on i-Math Investigations, and scroll down to the section for grades 3–5), all students can enjoy Exploring Geometric Solids and Their Properties, a five-part interactive geometry investigation. The activities, designed for students in grades 3–5, offer experiences in analyzing properties of two- and three-dimensional shapes and in developing mathematical arguments about geometric relationships. They also provide excellent opportunities for visualizing.

Students who need a greater challenge can be encouraged to look at the relationships among the faces, edges, and vertices of three-dimensional shapes using "Counting Parts of Solids." Looking for number patterns, they might discover Euler's formula (faces + vertices = edges + 2).

Students need a variety of experiences with three-dimensional shapes. Teachers often introduce these shapes and fail to allow ample time for exploring them. Continue these investigations by asking students to build 3-D shapes from specific sets of instructions rather

Leonhard Euler is credited with recognizing that the number of faces plus the number of vertices equals the number of edges plus 2. A biography of Euler, a worksheet to accompany the exploration with solids, and other activities related to Euler's contributions to mathematics can be found in Historical Connections in Mathematics *(Reimer 1992).*

than from actual models. For example, have the students build a shape that is made up of exactly six square faces, and ask the following questions:

- Can you make more than one model that fits this description?
- How many vertices and edges does the shape have?
- Now try making a figure from four equilateral triangles and a square. What do we call this figure? Count the vertices, edges, and faces.
- Given two rectangles, what kinds of other shapes can you add to make a three-dimensional figure? Make as many different figures as you can.

Searching for the Perfect Solids

Grades 4–5

Goals

- Discover the five perfect solids (also known as Platonic solids)
- Develop mathematical arguments to justify conclusions

Identify, compare, and analyze attributes of … three-dimensional shapes and develop vocabulary to describe the attributes

Make and test conjectures about geometric properties and relationships and develop logical arguments to justify conclusions

Build and draw geometric objects

Prior Knowledge

Students should be familiar with building and using three-dimensional objects. They should have facility with using the terms *faces*, *edges*, and *vertices* and understand how the terms are used in defining the characteristics of three-dimensional figures. They should also be familiar with the relationship of two-dimensional figures to three-dimensional shapes (e.g., a cube is made up of six faces that are squares). "Building Solids," the previous activity, can serve as a preparation for this activity.

Materials and Equipment

- Models of the five perfect solids plus a triangular prism and a square pyramid

  These models are commercially available or can be created from the patterns on the blackline masters "Patterns for the Perfect Solids" and "Patterns for Other Solids."
- "Sticks" (toothpicks, coffee stirrers, or such commercially available products as D-Stix) and "fasteners" (marshmallows, gumdrops, or clay) from which to construct solids
- A large recording space on which to keep track of properties and discoveries (chart paper, an overhead projector, or a classroom board) and writing tools

pp. 106, 108

Learning Environment

This activity is a combination of work in pairs and whole-class discussion. The "Engage" section describes a whole-class activity, and the "Explore" section involves work as partners and whole-class discussion.

Important Geometric Terms

Polygon, edge, octahedron, dodecahedron, polyhedron, vertex, cube, polygonal region, face, tetrahedron, icosahedron

Platonic solids: Three-dimensional figures composed of regular polygonal regions (all sides and angles equal in measure), with the

This activity has been adapted from "Why Are Some Solids Perfect?" (Lehrer and Curtis 2000), which is available on the CD-ROM that accompanies this book.

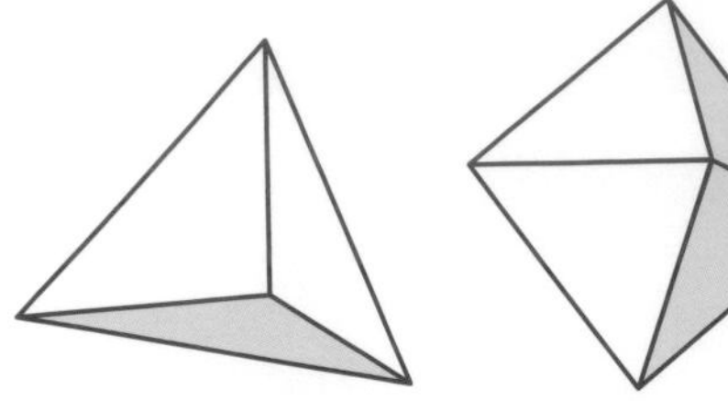

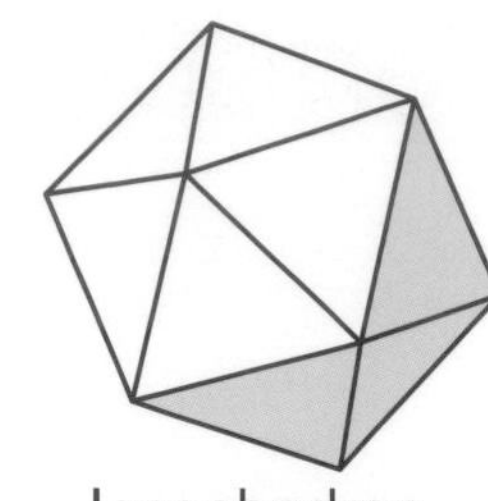

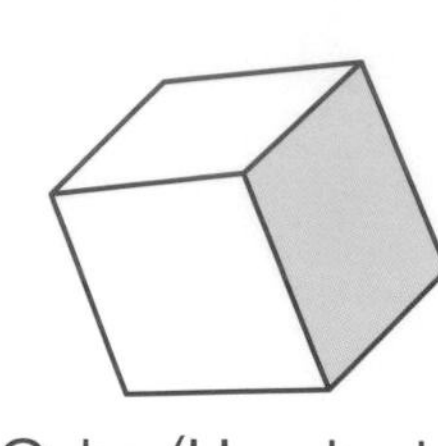

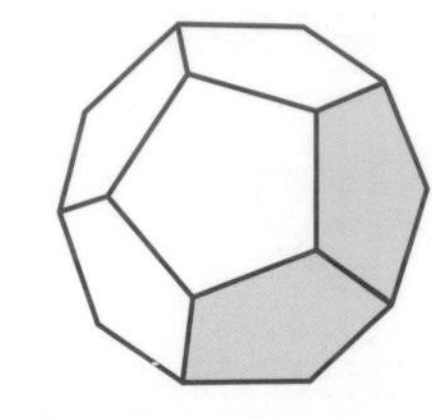

same number of faces meeting at each vertex. The faces can be equilateral triangles, as in the tetrahedron, the octahedron, and the icosahedron; squares, as in the cube; or pentagons, as in the dodecahedron.

Activity

Engage

To begin the lesson, show the class models of two perfect solids (a tetrahedron and a cube) and of two solids that are not perfect (a triangular prism and a square pyramid) and identify each model. Ask, “Looking at the examples that are perfect solids and those that are not, what do you notice?” Make a list of the properties the students identify. Someone might suggest that all the faces must be the same for a solid to be perfect. Someone else might think that a perfect solid always has three faces at each vertex. Ask the students to explain their reasoning. If the class agrees that an identified property makes sense according to its evaluation of these models, then write down the property. The students will test the properties as they create their own models. Tell them that there are a total of only five perfect solids. Their challenge is to find the remaining three and determine what makes a solid perfect.

The ancient Greeks, especially Plato, believed that these regular polyhedra represented the basis of the universe and the theory of matter. The tetrahedron represented fire; the cube, earth; the octahedron, air; the icosahedron, water; and the dodecahedron, the cosmos.

Only five arrangements of congruent, regular polygons can meet at a point and be folded to form a vertex of a solid (see fig. 1.14). Only the five solids resulting from folding those arrangements are perfect.

Fig. **1.14.**

The five arrangements of regular polygons that, when folded, form the perfect solids

Explore

Have the students work in pairs to make another perfect solid with “sticks” and “fasteners.” When a pair thinks it has succeeded, have the students share their model with the class and justify why they believe it is perfect. Either affirm or deny their conjecture and put the solid next to either the examples or the nonexamples. It is important not to tell the class why it does or does not belong. Use this opportunity to generate discussion by asking such questions as these:

- How has this new information changed your thinking about the list of possible properties?
- Should we add something to our properties list, take something away, or change something?

Have the students make and justify conclusions and change the list of properties accordingly. Continue to have the students present their "new" perfect solids. They will probably find several solids that fit the properties listed but are not perfect solids. They may, for example, have listed the following properties:

- Each face of the solid must be the same shape.
- Each face must be a shape that is equilateral.

Using these properties, the students might construct a polyhedron made up of rhombuses and think it is perfect. You may need to encourage the students to compare this solid to a cube:

- Look at each face. How is a face of the cube like a face of your new figure? How is it different?
- What might your observations tell you about a perfect solid?

The revelation that the sides of a figure can be the same length while the angles differ in measure can be new and exciting for students. They will then need to revisit their list of properties and begin to realize that their list is not incorrect but rather incomplete. This discovery will help them develop an appreciation of the power of properties working together, not in isolation, to create mathematical classifications.

Finally, finding the dodecahedron and icosahedron can be challenging, since they are harder to create with just sticks and fasteners. The students should first decide what shape each face must be. If they have not already discovered the limited possibilities for the faces (squares, equilateral triangles, or pentagons), encourage them to do so first. They can then make the faces out of oak tag and try taping them together to form solids. You may want to do this final model as a whole class activity, since it is time-consuming.

To summarize the activity and check for understanding, relate the list of defining properties to all five perfect solids and ask the students to point out how each property fits each of the solids.

All perfect solids are regular polyhedra; that is, each is made up of a single type of regular polygon, either an equilateral triangle, a square, or a regular pentagon. At each vertex, the same number of polygons intersects. In a cube, for example, three squares intersect at each vertex.

Assess

This lesson is exploratory, and students will have gained varying levels of new knowledge and different depths of understanding. A journal-writing activity will help you learn about the extent of their new knowledge. Ask the students to write a letter to a friend explaining what they have learned about perfect solids and exactly what makes a solid perfect. Encourage them to use diagrams, drawings, and nets (two-dimensional representations of three-dimensional figures) to visually enhance their explanations.

Extend

Connecting mathematics to other disciplines and to the world in which students live can extend this activity. For a language-arts connection, students might write a story about a wiggly octahedron that cannot keep its shape. They can gather information from the Internet and other resources about occupations that require a knowledge of how to make a cube or a tetrahedron and make a list of them. Two sites to help students get started are www.discoverengineering.org and www.manufacturingiscool.com. The students can research the origin of

Platonic solids and their importance to Greek mathematicians and philosophers, such as Plato.

The students can use the CD-ROM or visit the NCTM Illuminations Web site, illuminations.nctm.org (click on i-Math Investigations, and scroll down to the section for grades 3–5), to explore the interactive geometry investigation Exploring Geometric Solids and Their Properties. In the five activities found in this investigation, students can analyze properties of two- and three-dimensional shapes and develop mathematical arguments about their geometric relationships.

To further develop spatial visualization, you can ask the students to imagine what their figures would look like if they were "pulled apart." Have them create nets for each of the perfect solids and the nonexamples they have constructed. They can also explore these solids with the software program Shape Up! (Arita 1998), by Sunburst. In it, Plato's World allows students to manipulate the perfect solids to match shadows and nets with the correct figures. The students then color each face of the solid to match its corresponding face on the net.

Where to Go Next in Instruction?

These and similar explorations of two- and three-dimensional shapes and their properties and relationships lay a foundation for more-formal analyses of such concepts as congruence and similarity in higher grades. They also give students experience with important mathematical processes such as making and testing hypotheses and comparing and classifying objects. A good understanding of concepts of shape is important as students explore the essential ideas of location and position, which are examined in the next chapter.

GRADES 3–5

NAVIGATING *through* GEOMETRY

Chapter 2
Location

Students in prekindergarden–grade 2 come to understand locations in their world by developing skills that relate to direction, distance, and position in space. The language of navigation—for example, *up*, *down*, *around*, *in*, *into*, *on*, *left*, and *right*—is developed through play, physical experiences, literature, visual arts, visual reference points, and conversation. In addition, mapping experiences include locating objects and determining distances on number lines and graphs, using positive numbers and points of origin as reference points.

The first activity in this chapter, Find the Hidden Figure, helps students in grades 3–5 extend their understanding and language by interpreting directions for moving from one location to another. Beginning in the classroom, students should locate north, south, east, and west. Experiences should focus on giving such directions for moving to a specific spot in the classroom as "Take three steps north" or "Take five steps south." Students should extend their understandings through map work that includes multiple locations, positions, and distances in the school, then in the neighborhood and the town.

By grade 5, students' activities should include experiences that lead to their understanding of the coordinate plane and the origin (0, 0)—the starting point for finding and labeling all other points. In the next activity, *X*s and *O*s, students use Cartesian coordinates, or ordered pairs (x, y), to locate points on the horizontal and vertical axes and use the positive and negative integers in the four quadrants. This work should lead to plotting points on grid paper to make maps, to read and find locations, and to predict distances. The activity Can They Be the Same? uses coordinates to explore the vertices of similar and congruent polygons as they flip, slide, and turn in space.

Many connections to, and applications of, coordinate points are found in the world of science, social studies, and geography. In these disciplines, students must be able to use and understand coordinate systems to determine locations, distances, and positions.

Find the Hidden Figure

Grades 3–4

Goals

- Understand and use the compass directions north, south, east, and west
- Compose specific multiple-step directions to lead to a location
- Follow multiple-step directions to reach a location successfully

Describe location and movement using common language and geometric vocabulary

Prior Knowledge

Students should be able to use such positional directions as forward, backward, up, down, left, and right as they describe and interpret direction and distance in navigating space.

Materials and Equipment

- A compass, if available, to establish the directions north, south, east, and west
- Signs reading "North," "South," "East," and "West"
- Construction paper to make a geometric shape (e.g., a triangle) to hide
- Masking tape

Learning Environment

Students work in pairs.

Important Geometric Terms

North, south, east, west

Activity

Engage

Ask the students to describe experiences in which they or their parents have needed directions to find a place, perhaps when walking in the school or the neighborhood or when traveling in the car. If the children mention important positional words, list them on the board. You may need to model directions using at least two positional words. For example:

> If I were to give you directions to find the nurse's office, I would tell you to face west as you leave the room, follow the hallway to the end, and turn south. At the corner, turn west and walk to the next corner, where you turn north. Walk to the nurse's office door, where you turn east to enter her office.

Be sure to differentiate these *fixed directions* established with the use of a compass from the *relative directions* (e.g., right, left, forward, backward). Clarifying the terms will avoid misunderstandings and inaccuracies.

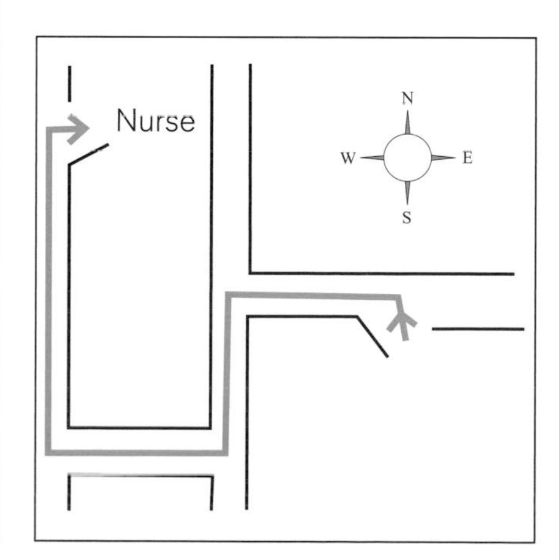

Introduce the four major compass directions: north, south, east, and west. Show, if possible, how you used a compass to determine north in your classroom. Place the sign "North" in an appropriate spot in your room. Ask, "What other direction could we use to describe another location in the room?" Elicit the remaining compass directions. Draw a

If your science curriculum includes lessons on magnets and compasses, help the children make the necessary connections to the fixed direcions.

circle on the board and label the directions to show the relationship of one to another. Ask the students to use the circle to decide where to post the remaining signs to further display the relationships.

To reinforce the relationships, call on individual students to move to certain locations. You might say, "Stand at the north wall of this room" or "Find the east wall in our classroom and stand below the sign." You can also have the children tell you what they see in the north side of the room.

Tell the students that you have hidden a triangle in the room and that you will lead one student to the triangle by reading a set of directions. Mark the starting point with a masking-tape X. Make a list of about five instructions that direct the student to the hidden triangle. Be sure to include turns and the number of steps toward north, south, east, and west. So students can practice taking steps of a uniform size, place at various spots on the floor of your classroom strips of masking tape cut in lengths equivalent to the size of the estimated "average stride" of the students. The following is an example of a set of directions:

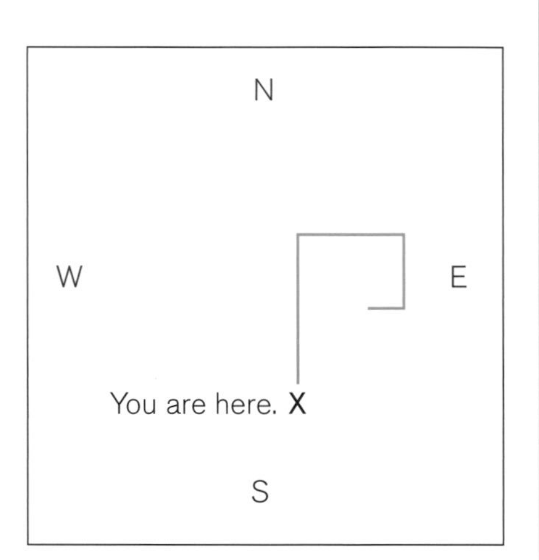

1. From the starting X, take four steps north.
2. Turn to the east.
3. Take three steps to the east.
4. Turn south and take two steps.
5. Turn west and take one step.
6. Look for the triangle.

Select a student and give him the directions. Follow the student's discovery of the triangle with a discussion, asking questions such as the following:

- What helped Julio find the triangle? (the directions)
- Could Julio have found the triangle without my using the terms *north*, *south*, *east*, and *west?* (Possibly, by using the terms *right* and *left*)
- Are there any other ways Julio could have found the triangle?
- Was I successful in writing my directions? How do you know? (Julio found the triangle.)
- Why do you think we have established the directions north, south, east, and west? (People all over the world use words in English or other languages to describe directions. Maps are made and instructions for finding locations are given using these terms. Also, the fixed directions do not depend on a person's orientation, as directions such as *right* and *left* do.)

Explore

Have the students pair up to write a set of directions that will lead other students to find a geometric object they have hidden. Remind the students that if they write detailed directions, they should be able to lead another person to their hidden geometric shape. They will be deemed successful if their shape is found!

This task can be presented to the students as the drawing of a treasure map. A blank map of the classroom will help them write directions and draw the intended path.

Assess

During informal observations, note the processes students use to compose their directions. Having another student follow the directions to find the hidden shape will give you information about the students' abilities to use the compass directions and to compose and follow specific directions. Each student should have a chance to try to follow someone else's directions.

Extend

You could move this activity to a larger area (the gymnasium or the playground), where you would need to establish the compass directions. The students might then write a set of directions to guide another student toward a specific location. Another variation is to translate this activity to a paper representation: Using graph paper with a compass key, a student can create a set of specific instructions for another student to move a marker toward a "magic spot" or a "hidden treasure."

Connect this activity to map reading in social studies. Point out the compass symbols on a real map. Ask the students to find several locations:

- Name the province that is north of North Dakota.
- What state is west of Connecticut?
- What country is immediately south of Texas?

Xs and Os

Grades 3–5

Make and use coordinate systems to specify locations and to describe paths

Goals

- Locate points on a rectangular coordinate plane using ordered pairs
- Use the point of origin (0, 0) as a point of reference
- Understand and use positive and negative integers to identify points in four quadrants

Prior Knowledge

The students should be familiar with number lines and moving forward and backward along them. They should know *left* and *right* and be able to name positive and negative numbers. (Thermometer readings in weather reports are useful in establishing the vocabulary for positive and negative numbers.) Through graphing activities, the students should have had multiple experiences using grids. It would also be helpful for them to have played ticktacktoe, or at least be familiar with the game. They should have practiced writing explicit instructions. As an example, you might perform an activity (e.g., get out from under a desk, make a peanut-butter-and-jelly sandwich, go to a closet and open the door to remove a designated item) at the direction of a student.

Materials and Equipment

pp. 109–12

- A copy of blackline masters "Coordinate Grids, A–D" for each pair of students
- An overhead projector with a transparency copy of each coordinate grid
- Transparency markers and crayons or colored pencils (one red, one blue)

Learning Environment

For the modeling activity, divide the class into two groups. Designate one group the blue team and the other the red team. During the exploration time, students work in pairs.

Important Geometric Terms

Coordinate points, ordered pair, x-axis, y-axis, point of origin, horizontal, vertical, diagonal

Activity

Engage

Tell the students that they are going to play a game similar to ticktacktoe. Half the class, the red team, whose symbol is *X*, will play against the other half, the blue team, whose symbol is *O*. Display "Coordinate Grid A" on the overhead projector, and discuss how the game is to be played: You will call one of the teams. One person from the team will tell you where to place an X or O by naming an ordered pair. All the students' responses must be given in terms of ordered pairs. Explain that the first number called locates the position along the x-axis and the second number of the ordered pair locates the position along the y-axis. Be sure that the X or the O is placed on the intersection of the grid lines.

When you play the game, emphasize the point of origin as you count aloud from the origin to represent the first number in the ordered pair. As the students give you successive coordinate points, plot them on the graph. The first team to have four *X*s or four *O*s in a row—horizontal, vertical, or diagonal—will be the winner.

Allow time for the teams to discuss who their spokespersons will be and to talk about their strategy for the placement of their symbol (X or O) on the first move. You might post the following questions in the classroom to stimulate discussion and help the students develop problem-solving strategies. Then repeat the questions after the first game has been completed.

- What are some strategies that will help you win this game?
- Is there a pattern?
- Is it better to try to get your marks in a row or to try to block the other team?
- How will you play the game differently next time? (*Use this question in the discussion after the first game.*)

Play the game again on the same grid, recording the *X*s and *O*s exactly as the teams indicate. Remind the spokespersons to check with their teammates before calling out the next move. Allow time for consultation if necessary. Note the changes in the strategies that each team is implementing. Incorporate your observations into the discussion after the second game. Again discuss the questions in the list above. The students may share ideas such as "It is always better to begin with (2, 2) if you want to win" or "You need to get your first two in a row and then block the other team." You might want to try the game in this format once more or move on.

You may want to research games of other cultures with features similar to those of tacktacktoe. For example, "go" and "go-moku," played in Asian cultures, require placing five markers in a row.

Explore

Distribute "Coordinate Grid A," pair the students to play the game, and observe them as they play against each other. Listen to their conversations, and challenge their thinking with questions similar to those posed earlier. Ask additional questions such as these to help them think about their problem-solving strategies:

- When did you know you would win?
- What does it take to win this game?
- Would the game be harder or easier if we added more rows? Why?

Assess

Have the students write in their journals, using prompts such as these:

- How do you think you can win the X and O game?
- Write down directions for another student to follow in placing your symbol. Be sure to be specific so that your X or O is placed where you want it to be.

Sharing some of the students' journal responses will help other students clarify their understanding and think about new strategies.

In the response to the second prompt, look for the ordered pair to be mentioned.

Extend

Adding quadrants to the coordinate grid, as in the remaining blackline masters for this activity (see the reduced copies in fig. 2.1), extends children's thinking and moves the students into using negative numbers. (It might be necessary to review the use of negative numbers prior to introducing the expanded graphs.)

Another activity involving ordered pairs to locate points requires lining a large area—perhaps a playground or gymnasium—with grid lines to represent the horizontal and vertical lines of a graph. Establish the origin, and then call a child's name, assign the child an ordered pair, and have him or her move to the assigned location on the grid. Assign a point on the grid to each child. You can then have the student at one point—for example, (2, 3)—exchange places with the student at another point—for instance, (4, 5). You can create many extensions by having the children assign locations to classmates or design games to use with the large grid.

Many additional games are available to reinforce the concept of locating points on a rectangular coordinate plane. For example, Hurkle (Stenmark, Thompson, and Cossey 1986), Where's the Rectangle? (Downie, Slesnick, and Stenmark 1981), Coordinate Tic-Tac-Toe

Fig. **2.1.**
Expanded coordinate grids

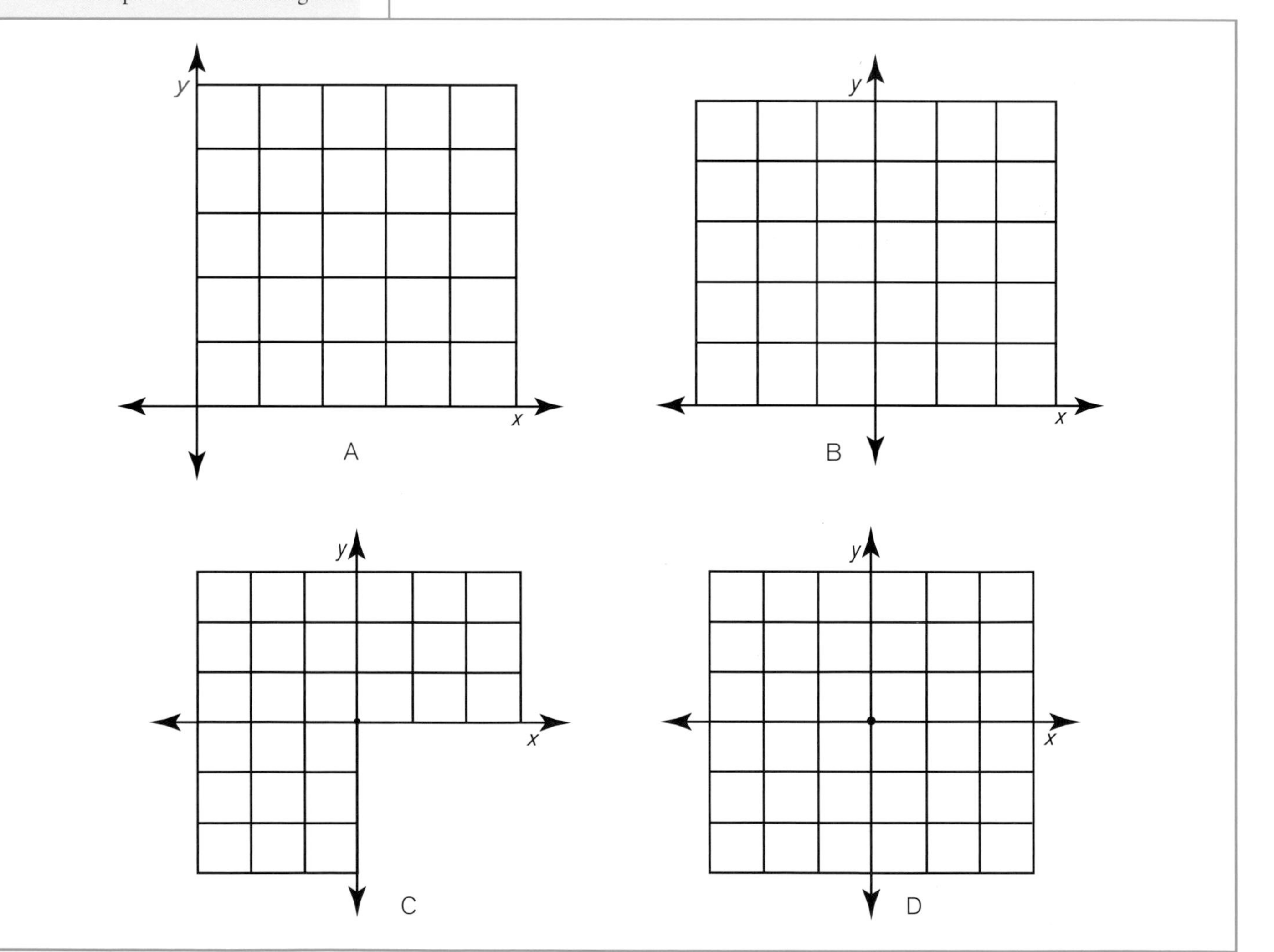

(Stenmark, Thompson, and Cossey 1986), and Coordinates I and Coordinates II (Stenmark, Thompson, and Cossey 1986).

Identifying points on a coordinate grid is important in understanding how the coordinate system works and in constructing simple line graphs to display data or to plot points. These skills can be used to examine algebraic functions and relationships.

The skills developed in this activity can be applied to interpreting latitude and longitude in map reading in social studies and to plotting points to represent data collected and recorded during science experiments. Consider the use of the coordinate plane when extending ideas related to symmetry, reflections, and spatial sense.

Can They Be the Same?

Grade 5

"An understanding of congruence and similarity will develop as students explore shapes that in some way look alike. They should come to understand congruent shapes as those that exactly match and similar shapes as those that are related by 'magnifying' or 'shrinking.'"
(NCTM 2000, p. 166)

Goals

- Explore similar shapes
- Develop an understanding of similarity
- Test for similarity of shapes using the coordinate grid
- Explore the effects of magnifying or shrinking a shape

Prior Knowledge

Students should have an intuitive notion of similarity. They should also have a beginning understanding of coordinate points developed through the use of number lines.

Materials and Equipment

pp. 113–15

- A copy of the blackline master "Similar and Nonsimilar Shapes" for each student
- A copy of the quarter-inch and half-inch "Grid Paper" blackline masters for each student
- An overhead projector
- A transparency of each of the "Grid Paper" blackline masters for use on the overhead projector

Learning Environment

The students work in small groups or individually.

Important Geometric Terms

Nonsimilar

Similar: Two shapes are similar when their corresponding angles are congruent and their corresponding sides are in proportion. Similar figures have the same shape but may not have the same size.

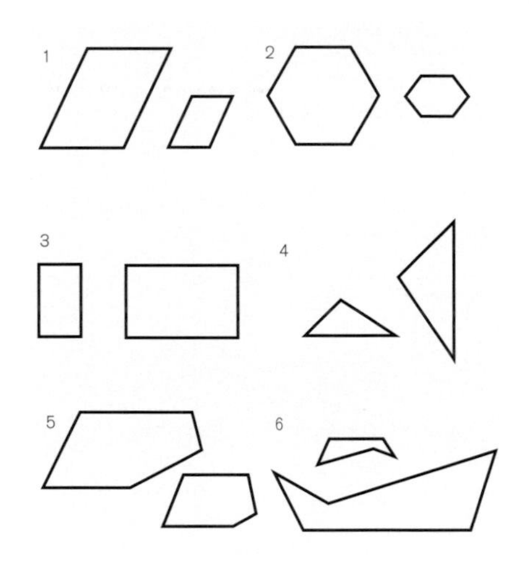

Activity

Engage

Present students with the blackline master "Similar and Nonsimilar Shapes." Ask questions like the following about the six pairs of shapes:

- Which pairs of shapes are similar?
- What makes them similar? Describe how they are similar.
- How can you show that they are similar?
- How are the similar shapes alike? How are they different?

The students might tell you that "they look alike." Elicit additional details about the shape, the length of the corresponding sides, the size of the angles, and the number of lines.

Keep a list on the board or on chart paper to record the students' ideas as they think aloud about similarity. Use a similar line of questioning to examine the nonsimilar shapes. List the responses the children give you for the nonsimilar shapes. Through discussions and more examples, which you can produce, lead the students to understand that similar shapes have the same shape but not necessarily sides of the same size.

Another to way to help students explore similarity is to use coordinate grids. Display a drawing on the overhead projector, first using the half-inch transparency grid. Then display a similar drawing on the quarter-inch transparency grid. On each transparency, list the coordinate points used to draw the picture. (See fig. 2.2.)

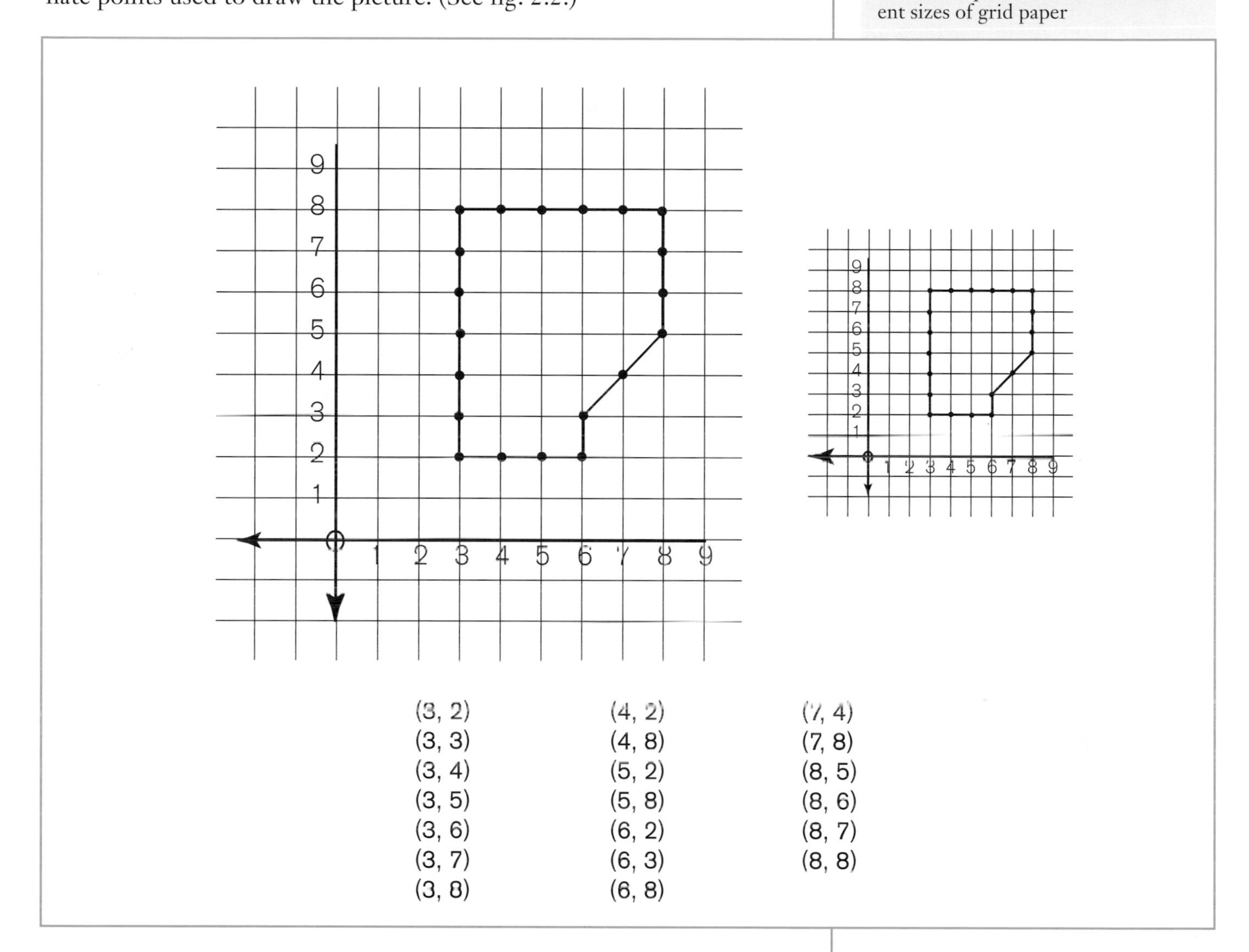

Fig. **2.2.**

Similar shapes drawn on two different sizes of grid paper

Discuss with the students how using the coordinate grid could illustrate the similarity of two figures. Elicit from the students that the coordinate points are the same for each figure and that only the size of the grids and the size of the figure have changed.

Explore

In the task What shape will this make? (adapted from "Find the Picture" [1986]), the students create a shape by plotting points on a coordinate grid. Give the children the quarter-inch and the half-inch grid paper for this activity. Using both sizes of grid paper will help the students make a connection to the ideas of magnification and shrinking. Tell them to plot the following points on both coordinate grids: (2, 3), (2, 4), (5, 4), (5, 6), (7, 6), (7, 20), (9, 4), (9, 3), (13, 3), (11, 1), (2, 1), (1, 3). Ask the students to share their ideas about the similarity of the

The emphasis is on similarity as it relates to dilation (magnification or shrinking) of the shapes. The coordinate points are added to help the students develop an understanding of similarity.

two pictures (a boat with a mast). Use the questions in the "Engage" section of this activity for discussion.

Ask the children to draw on one-half- and one-quarter-inch grid paper two sets of similar shapes of their own design and then list the coordinate points of the vertices. After the shapes have been created, the students exchange them to have another student determine if the shapes are similar by checking that the coordinate points are the same on the two figures.

Assess

Informal observations of process and dialogue will give you valuable information about the students' understanding of similarity. Listen to students giving "evidence" to support the similarity or nonsimilarity of shapes. The evidence might include comments such as "All the coordinate points are the same in the list of coordinate points" or "Each side of the shape on the quarter-inch grid paper seems to be one-half the size of the side of the shape on the half-inch grid paper. We can test this out by measuring to see." In addition, using different sizes of coordinate graph paper to have students create similar shapes can be used as an assessment tool when coupled with writing journals entries. Have the students describe what makes two shapes similar. Ask how they know if two shapes are similar. You can also have one student draw a picture on grid paper and list the coordinate points for the figure. Then give just the list of coordinate points to another student to create a similar shape on graph paper of a different size.

Extend

Take the students outside to look for examples of similar shapes in nature (leaves from trees, petals of flowers, etc.) and have them write how the shapes are similar.

Other extensions can be made to map work and scale drawings. For instance, when the class is locating points on maps in social studies activities, ask the students if they see any connections to their work with coordinate points in graphing. They should recognize that locations on a map are identified by intersections of lines of longitude and latitude. The points of intersection are similar to coordinate points.

When increasing the scale on scale drawings, using a grid and coordinate points helps students see the effects of changing the scale: doubling the scale, for example, reduces the length of each side of the scaled figure by one-half.

Where to Go Next in Instruction?

This chapter on location emphasizes developing the skills needed to understand and use coordinate points on a grid to form an object or determine the location of an object. In addition, the activities provide experiences in using the positive and negative integers. Working with coordinate points, or ordered pairs, students can explore similar and congruent shapes through magnifying or shrinking. The skills developed in studying location can be applied to exploring congruent polygons that have been transformed in space, through either a flip, a slide, or a turn. These transformations are the subject of the next chapter.

GRADES 3–5

NAVIGATING *through* GEOMETRY

Chapter 3
Transformations

Studying transformation geometry gives students experience with how the orientation of a shape changes as it moves in space. *Principles and Standards for School Mathematics* (National Council of Teaches of Mathematics [NCTM] 2000) suggests that students in grades 3–5 move beyond the initial exploration of slides, flips, and turns introduced in the lower primary grades and that they use more-sophisticated vocabulary to describe the results of transformations on shapes. For example, statements like "The square was rotated 90 degrees" or "The triangle was reflected horizontally" should be encouraged. Students in grades 3–5 should be able to visualize and predict the results of transformations and use motions or a series of motions to show that two shapes are congruent. As students develop confidence and proficiency in working with transformations, the terms *translation*, *reflection*, and *rotation* should be introduced.

Line and rotational symmetry should also be investigated in grades 3–5. *Principles and Standards for School Mathematics* emphasizes that symmetry should be viewed as an aid in describing the geometric properties of shapes. For instance, students may discover that an equilateral triangle has three lines of symmetry whereas a nonsquare rectangle has only two. Students need to explore and design figures with more than one line of symmetry and use precise and sophisticated language for rotations and angles when describing rotational symmetry. Hands-on activities that incorporate paper folding, reflective devices, and tracing are vital in developing understanding.

The activities in this chapter are designed to develop students' abilities to apply transformations and to use symmetry to analyze mathematical situations, as outlined in *Principles and Standards*. These activities afford students opportunities to gain deeper understandings as they move from working with simple geometric shapes to working with more-complex figures. Attention is given to real-world applications of these topics to encourage students to connect the mathematics they are studying to their world.

The activities Patchwork Symmetry and Symmetry Detectives—Learn the Secret Code! focus on identifying and describing lines of symmetry. In the first, students use reflective devices to discover lines of symmetry in pattern blocks and extend their understanding by combining several pattern blocks to create patchwork-quilt squares that have line symmetry. In the second, students identify lines of symmetry in letters of the alphabet and then apply this knowledge by writing and deciphering words in "secret" symmetry code. Going Logo for Symmetry! is an in-depth investigation of rotational symmetry in which students explore a variety of logos and discover ways to identify and describe the angle of rotation in figures that have rotational symmetry.

The remaining three activities offer opportunities to apply translations (slides), reflections (flips), and rotations (turns). In Tetrominoes Cover-Up, students use these motions to show whether two shapes are congruent as they create a set of tetrominoes (figures consisting of four squares). Then they play a game using translations, reflections, and rotations to manipulate tetromino shapes in order to completely cover a rectangular region. As the students play the game, they are encouraged to visualize and predict how a given tetromino shape can be moved to fit best within the rectangular region. In Motion Commotion, students move a shape through a series of transformations, record the moves on a paper strip, and then challenge classmates to replicate the series of moves. In Zany Tessellations, students discover how transformations are used to create tessellations. A connection is made to art by having students explore the tessellations found in the art of M. C. Escher. Students are given opportunities to create their own tessellations by hand and also with computer software.

As students explore the challenging and creative activities in this chapter, they gain a deeper understanding of transformation geometry and develop more-precise and more-sophisticated language to describe how the orientation of a shape changes as it moves. Making connections to real-world applications of these concepts fosters students' appreciation of the balance and forms that surround them.

Patchwork Symmetry

Grades 3–4

Goals

- Identify lines of symmetry in various pattern blocks
- Use pattern blocks to design various patchwork-quilt squares that have line symmetry

Identify and describe line … symmetry in two- and three-dimensional shapes and designs.

Prior Knowledge

Students should have had experience identifying lines of symmetry in basic geometric shapes through paper-folding activities in the primary grades.

Materials and Equipment

- An overhead projector and a transparency, overhead pattern blocks, and washable transparency markers
- Pencils, markers or crayons, and glue or glue sticks
- For each student, a set of at least twenty pattern blocks and several sets of paper pattern-block shapes (Pattern-block templates are available on the CD-ROM that accompanies this book)
- Two six-inch "quilt squares" cut from white paper for each student (A "Quilt-Patch Work Space" template is available on the CD-ROM that accompanies this book.)
- A small, handheld mirror or plastic mylar mirror for each student
- *The Patchwork Quilt*, by Valerie Flournoy (1985), or *Sam Johnson and the Blue Ribbon Quilt*, by Lisa Ernst (1983)

Important Geometric Terms

Line of symmetry: A line that divides a figure into two halves such that the halves are mirror images of each other.

Learning Environment

Students work in pairs for the entire lesson. However, individual students make their own patchwork-quilt square for the class quilt.

Activity

Engage

Read *The Patchwork Quilt* or *Sam Johnson and the Blue Ribbon Quilt* to stimulate students' interest in making their own patchwork-quilt squares. Inform the students that they will be designing and making their own squares, which will be special because the designs they make will have line symmetry.

The students can also design symmetrical patterns using the Pattern Patch applet on the CD-ROM.

Hand out approximately twenty pattern blocks, a mirror, and a blank

---

This activity has been adapted from *Twenty Thinking Questions for Pattern Blocks, Grades 3–6* (Walker, Reak, and Stewart 1995b).

Blue rhombus:
two lines of symmetry
Green triangle:
three lines of symmetry
Orange square:
four lines of symmetry

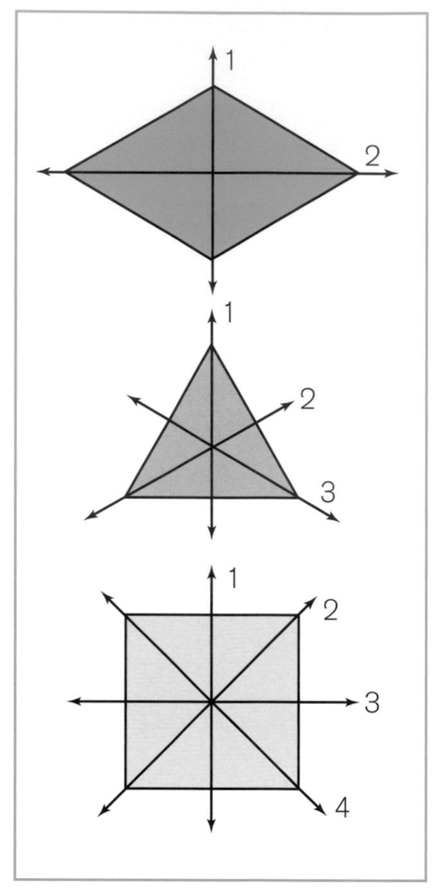

piece of paper to each student. Place the red trapezoid pattern block on the overhead transparency and trace around the edges. Have the students do the same on their paper. Ask them to find a line that divides the pattern block in half so that the two halves are mirror images of each other. Have a volunteer draw such a line on the transparency. Tell the class that this line is known as the *line of symmetry*. Model for students how to verify a line of symmetry by placing a hand mirror on the line of symmetry. The image reflected in the mirror should coincide with the image behind the mirror. The reflected image together with the image in front of the mirror should create the entire original figure.

Ask the students if the trapezoid has any other lines of symmetry. When the students are convinced that there is only one line of symmetry, use the same procedure to test the blue rhombus, the green triangle, and the orange square pattern blocks for lines of symmetry.

Next, place a red trapezoid and two blue rhombuses on the overhead projector as shown in figure 3.1. Trace along the edges of the design. Have the students do the same on their paper. Ask them to find a line of symmetry in the design. Have a volunteer identify the line of symmetry on the overhead projector. Then have the students verify the line of symmetry on their papers using a mirror. Ask them to try to find other lines of symmetry. Discuss the following:

- When does a design have a line of symmetry? (When the line divides the design into two halves so that each half is a mirror image of the other.)
- Is it possible to move the rhombuses to create a different design that has line symmetry?

(Some possible solutions are shown in figure 3.2.)

Explore

Each student builds a patchwork-quilt square for a class symmetry quilt. Hand out two paper "quilt squares" to each student. Instruct the students to use any ten pattern blocks to make a patchwork-quilt-square design that has exactly one line of symmetry and fits within the six-inch-by-six-inch square. When they have completed their designs, have them make two copies of their quilt square by gluing paper pattern blocks onto the blank quilt squares. Have the students use a marker or a pencil to draw the line of symmetry on the front of one patchwork-quilt

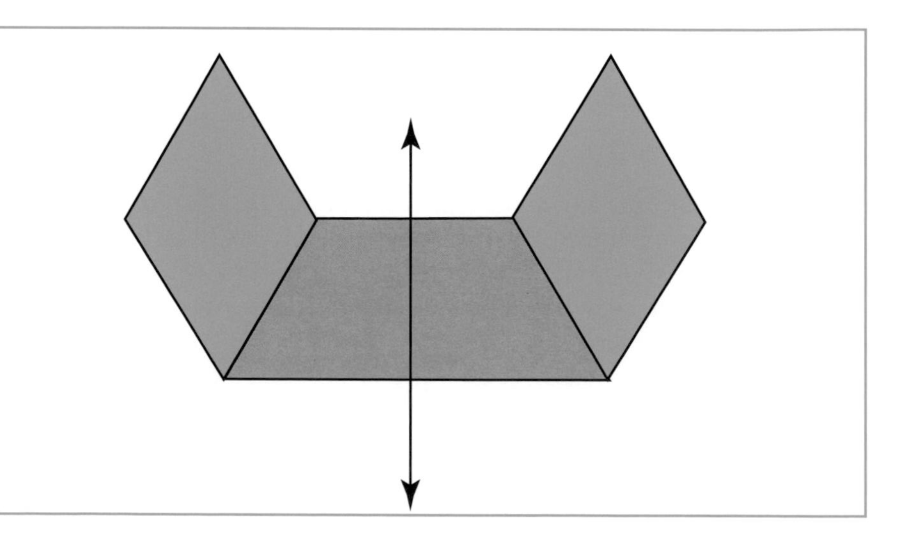

Fig. **3.1.**
A design made from a trapezoid and two rhombuses

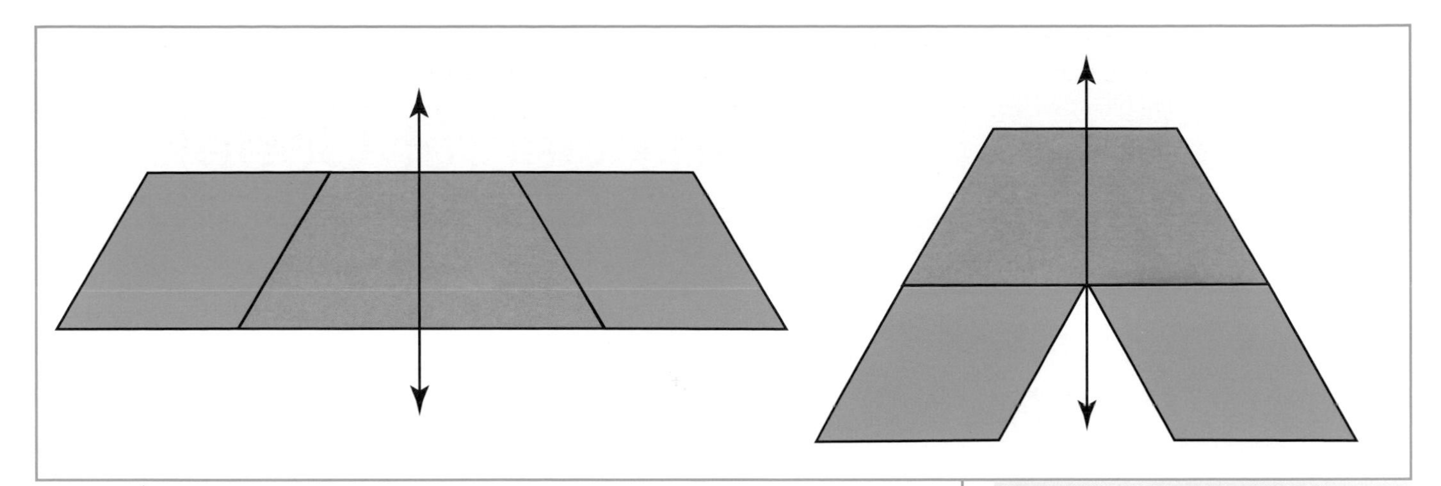

Fig. **3.2.**

Two alternative designs made by moving the rhombuses

design and write on the back how they know the line they drew is a line of symmetry. The students then trade the unmarked copy of their quilt design with a partner to verify the line of symmetry. Later they can use that copy to construct a class "symmetry patchwork quilt" to be displayed on a bulletin board in the classroom.

Assess

Assessment occurs primarily through observation and listening as children trade designs and verify lines of symmetry. Have the whole class share and discuss their work. Pay close attention to students' explanations of the lines of symmetry in their designs. Analyzing students' work will give you insights into their level of understanding.

Extend

Have the students look for other objects at school or at home that have line symmetry and identify the lines of symmetry, or have them use a larger number of pattern blocks to create more complex quilt designs that have two or more lines of symmetry. Encourage the students to browse quilting books to search for quilt patterns or patchwork-quilt squares that have line symmetry. *One Hundred One Full-Size Quilt Blocks and Borders* (Dahlstrom 1998) and *The It's Okay If You Sit on My Quilt Book*, by Mary Ellen Hopkins (1989), are some possible resources. The students who are confident with line symmetry may want to explore rotational symmetry with quilt designs.

A design or figure has *rotational symmetry* when it can be rotated about a point less than a full turn (360°) so that the rotated design or figure coincides with the original figure. The figure below has 60-degree rotational symmetry.

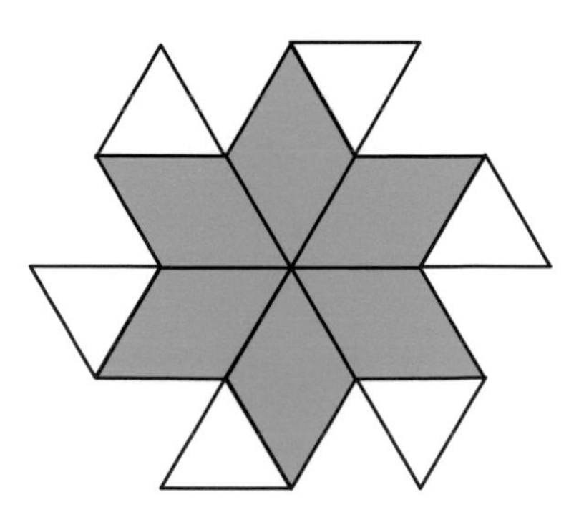

Symmetry Detectives–Learn the Secret Code!

Grades 4–5

Identify and describe line … symmetry in two- and three-dimensional shapes and designs

Goals

- Explore lines of symmetry in simple figures and geometric shapes
- Identify the lines of symmetry in letters of the alphabet
- Identify objects in the real world that have line symmetry

Prior Knowledge

Students should be able to use paper folding or cutting to determine if basic geometric shapes have line symmetry.

Materials and Equipment

pp. 116, 117

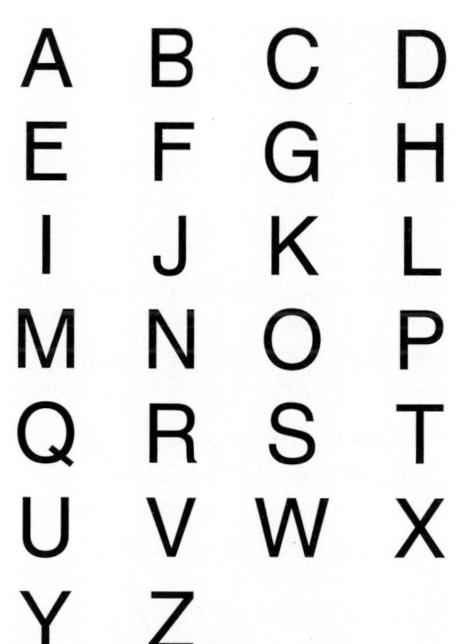

- A copy of the blackline master "Alphabet Symmetry" for each student
- A copy of the blackline master "Alphabet-Symmetry Chart" for each student
- A copy of the paper strip with figures at the bottom of the "Alphabet-Symmetry Chart" for each student
- A small mirror or Mira (or similar reflective device) for each student
- Pencils, scissors, and glue or glue sticks
- Old magazines and newspapers
- An overhead projector and transparency marking pens
- Students' mathematics journals
- An overhead transparency of "Alphabet Symmetry"
- An overhead transparency of the paper strip from the bottom of "Alphabet-Symmetry Chart"

Learning Environment

Students work in pairs to carry out all activities.

Important Geometric Terms

Line symmetry, line of symmetry

Activity

Engage

Tell the students that they will become detectives and look for lines of symmetry in various figures and geometric shapes. They will use a mirror or other reflective device to test for lines of symmetry. To test

---

Part of this activity has been adapted from Symmetry in the Alphabet, activity 3 from Symmetry and Tessellations (Jill Britton, www.camosun.bc.ca/~jbritton/jsbsymteslk).

for line symmetry, place the reflective device on a figure and move it around until half of the figure is reflected. This reflected image should coincide with the portion of the figure behind the reflective device. The line formed where the edge of the reflective device touches the figure is know as the *line of symmetry*. The images on either side of the line of symmetry should be mirror images.

Hand out copies of the paper strip. Use the rectangle on it to model this exploration. Ask the children to predict whether the rectangle has a line of symmetry. Then allow them to use the reflective devices to test for lines of symmetry. Any students who are having difficulty using the reflective devices could fold paper cutouts of the figures to test for line symmetry. To use the folding method, try to fold the figure in half so that the two resulting halves are congruent mirror images and coincide perfectly, one on top of the other. When such a fold line can be found, the fold represents a line of symmetry.

Place the transparency of the paper strip on the overhead projector, and have a volunteer draw a line of symmetry on the rectangle. Ask the following questions:

- How do we know this is a line of symmetry?
- Is there another possible line of symmetry?

Explain that although many figures have one or more lines of symmetry, some do not have any lines of symmetry. When the class has discovered all possible lines of symmetry for the rectangle, allow the students individually to test the remaining figures for lines of symmetry. (See the solutions to the paper-strip figures in the appendix.) When they have completed the activity, encourage them to share their discoveries.

Explore

Distribute a copy of "Alphabet Symmetry" and "Alphabet-Symmetry Chart" to each student. Tell the students that while working with a partner, they will continue their detective work by finding lines of symmetry in the letters of the alphabet. Model the activity using the **A** on the activity sheet. Ask the students to predict whether the **A** has a line of symmetry. Then allow them to use mirrors to check their predictions. Have a volunteer draw a line of symmetry on an overhead transparency on top of the **A**. Then ask the students to check for other possible lines of symmetry for **A**. When they have done so, have them count the lines of symmetry and write the letter in the appropriate column on the "Alphabet-Symmetry Chart." Allow the students time to explore the remaining letters of the alphabet for lines of symmetry and complete the two activity sheets. When they have done so, encourage them to share their discoveries. (The solutions appear in the appendix.) Discussion questions might include the following:

- Which letters have only one line of symmetry? (**A, B, C, D, H, M, T, U, V, W,** and **Y**)
- Why? (I can find only one line that divides the letter into two halves so that the two halves are mirror images of each other.)
- Which letters have no lines of symmetry? (E, **F, G, J, K, L, N, P, Q, R, S,** and **Z**)
- Why? (I cannot find a line that divides the letter into two halves so that the two halves are mirror images of each other.)

A Mira allows students to see both the original figure and the reflected image at the same time, making it easier to tell when the reflection coincides with the original. Mirrors do not.

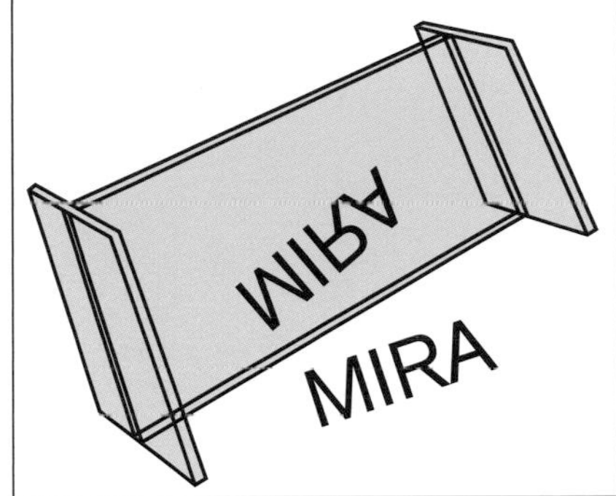

In this discussion, the term *vertical line* refers to a line that is parallel to the left or right edge of the activity sheet and *horizontal line* refers to a line that is parallel to the top or bottom edge of the sheet.

is "secret" symmetry code for MATH.

is "secret" symmetry code for SMILE.

- Which letters have two lines of symmetry? (**I, O,** and **X**)
- Which letters have more than two lines of symmetry? (none)
- Which letters have horizontal lines of symmetry? (**B, C, D, I, O,** and **X**)
- Which letters have vertical lines of symmetry? (**A, H, I, M, O, T, U, V, W, X,** and **Y**)
- If we used a different font or style to print these letters, would the line symmetries stay the same for each letter? (The lines of symmetry may be different.) Why? (For example, in the font on the "Alphabet Symmetry" activity sheet, the E does not have line symmetry because the upper section is slightly smaller than the lower section. In a different block style, both sections in the E could be the same size, and then E would have a horizontal line of symmetry.)

Next, tell the students that a "secret" symmetry code can be made using the line symmetry they discovered while working with the letters of the alphabet. Writing only one-half of a symmetrical letter creates a code that can be deciphered using a mirror or other reflective device. (Letters that do not have any lines of symmetry should be written as they normally appear.) Model on the board how to write a few letters in code.

Initially, have the students practice writing one word in "secret" symmetry code and give the coded word to their partner to decipher. In order to decipher the words, the students may need to use a mirror to find the line of symmetry for each letter. When the line of symmetry is found, the letter should appear in its original form. As the students gain confidence, they can write phrases or whole sentences in "secret" symmetry code for their partners to decipher.

Finally, discuss with students what other objects have line symmetry. For example, a "smile face" sticker or a window could have line symmetry.

Assess

Ask the students to find pictures in magazines or newspapers of objects that have no lines of symmetry, one line of symmetry, two lines of symmetry, and more than two lines of symmetry. Then have them trace or cut out the pictures, glue them in their journals, draw the appropriate lines of symmetry, and indicate the number of lines of symmetry.

Extend

Have the students explore a variety of fonts and determine which letters have line symmetry in the various fonts. Compare the results with those for the letters used in this lesson. The students can also explore lowercase letters. Have them search for words like **MOM** and **CODE** that have line symmetry or explore which numerals have line symmetry. The students could also keep a log for one week of objects they see at home and at school that have line symmetry. Encourage them to include both natural and manufactured objects.

The students should extend their explorations of symmetry by considering rotational symmetry. They could, for example, investigate which letters of the alphabet have rotational symmetry. See the next activity, Going Logo for Symmetry!, for a description of rotational symmetry.

Going Logo for Symmetry!

Grade 5

Goals

- Explore simple figures and geometric shapes for rotational symmetry
- Identify rotational symmetry in various corporate logos (trademarks)
- Design for a school club or activity a logo that has rotational symmetry

Identify and describe … rotational symmetry in two- and three-dimensional shapes and designs

Prior Knowledge

In the primary grades, students should have had experience with turning, or rotating, geometric shapes and observing the effect of rotation on the orientation of the shapes. They should also have had experience predicting the results of rotating a figure. Students should be able to distinguish between clockwise rotation (to the right) and counterclockwise rotation (to the left). It is recommended that students have experience in identifying line symmetry in objects. In addition, students should be familiar with such fractional parts of circular regions as one-half, one-fourth, one third, two thirds, and three-fourths. These fractions may be used to identify and describe various rotational symmetries found in shapes. Finally, students should be able to read angle measures using a protractor.

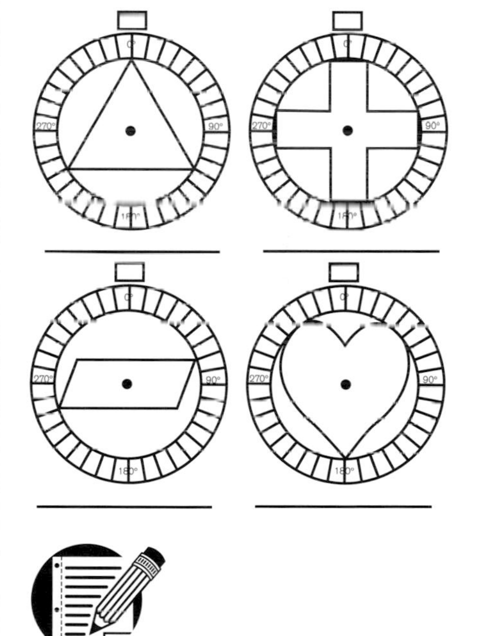

Materials and Equipment

- Overhead-transparency sheets cut into quarters, six quarter-sheets for each pair of students
- Washable transparency markers for the teacher and each pair of students
- An overhead projector
- Various rectangular shoeboxes with lids (one box for each pair of students)
- Various square gift boxes with lids (one box for each pair of students)
- A copy of the "Turn It Around" blackline master for each pair of students
- Old newspapers or magazines
- Scissors
- Students' mathematics journals

p. 118

Learning Environment

Students work in pairs to carry out all tasks.

This activity has been adapted from Renshaw (1986), which can be found on the CD-ROM that accompanies this book.

Important Geometric Terms

Rotational symmetry: A figure has rotational symmetry if it can be rotated less than a full turn (360°) about a fixed interior point and the rotated image matches, or coincides with, the original figure.

Center of rotation: A fixed point about which a figure is rotated

Point symmetry: Half-turn (180°) rotational symmetry

Activity

Engage

Distribute at least one rectangular and one square box and lid to each pair of students. Have the students mark an *X* near the midpoint of one edge of the rectangular lid and make a similar mark directly below it on the inside bottom of the box. Tell the students that the marks will help them remember the original position of the lid (see fig. 3.3). Ask, "Is there a way you could turn the lid and place it on the box so that the cover would fit properly, as it did in its original position?" Call for a volunteer to demonstrate one possible solution and describe how far the lid was turned (halfway around). Explain that when an object is rotated less than a full turn (360°) and the figure matches, or coincides with, the original figure, we say the figure has rotational symmetry. Therefore, the rectangular lid has rotational symmetry because when it was turned halfway around, the position of the lid coincided with the original position. The rectangular lid has half-turn (180°) rotational symmetry.

Fig. **3.3.**

A marked box and lid

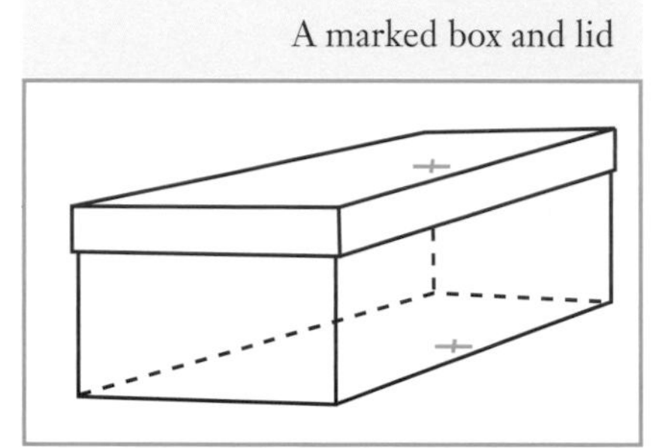

Since all figures coincide with their original position when they are rotated a full turn (360°), only rotations greater than 0 degrees and less than 360 degrees are considered when identifying rotational symmetry.

Ask the class to continue exploring with the rectangular lid to find other unique ways for the lid to fit on top of the box. When the students are convinced that the lid fits properly only in the original position and the half-turn position, have the students explore the square boxes and lids. Ask them the following questions:

- When does a figure have rotational symmetry? (A figure has rotational symmetry when it is rotated less than a full turn and the rotated image coincides with the original figure.)
- Does rotating the rectangular and square lids counterclockwise rather than clockwise make a difference in rotational symmetry? (No, the rotational symmetry will be the same.)

The square lid has three unique positions in addition to the original one that allow the cover to fit on top of the box as it did in the original position. They are the quarter turn (90°), the half turn (180°), and the three-quarters turn (270°). The angle describes how far the square lid is rotated from the original position in order to coincide with the top of the box; it is referred to as the *angle of rotation*.

Angle measures such as 270° may be challenging for some students. See figure 3.4.

Explore

Tell the students they will investigate some shapes to determine whether they have rotational symmetry. Hand out a copy of "Turn It Around" and six overhead transparency pieces to each pair of students. Model the task on the overhead projector, using the equilateral triangle as an example. Explain that it was easy to pick up the box lids and turn them to check for rotational symmetry but that when working with a figure that is drawn on paper, they must use a different approach to check for rotational symmetry. Ask the students for suggestions. Guide them in discovering that tracing the triangle and then rotating the traced image on top of the original figure is one way to check for rotational symmetry.

First, ask the students to predict whether the equilateral triangle has rotational symmetry. Then ask them to trace the triangle on an overhead transparency and mark their initials in the box above the figure to help them remember the original orientation of the traced image. Next, draw the students' attention to the point in the center of the equilateral triangle. Explain that the point located in the center of a figure is known as the *center of rotation* and that it remains fixed when the traced image is rotated on top of the original figure.

With the transparency image overlaid on top of the original triangle, have one student in each pair place a pencil point on the center of rotation and hold the activity sheet in place with his or her free hand. The partner then slowly turns the transparency clockwise. As the transparency is turned, the students should note when the traced figure coincides with the original figure before a full turn (360°) has been completed. When such a match occurs, rotational symmetry has been demonstrated.

Fig. **3.4.**

The lid of a square box rotated 90°, 180°, and 270°

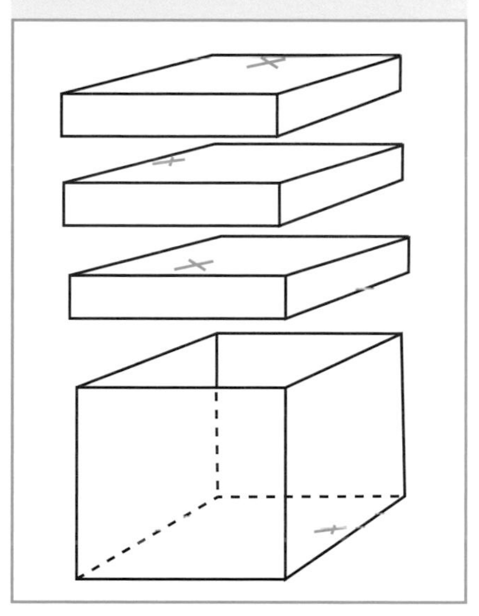

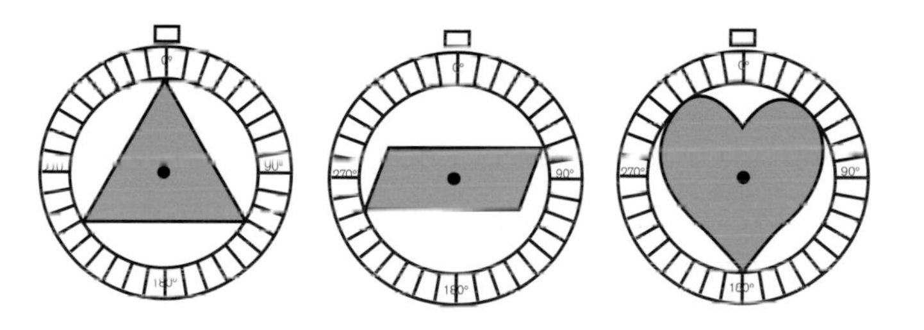

Figures have been drawn inside six of the full-circle protractors on the blackline master. The protractors can help the students discover the angles of rotation.

Direct the students to record on the original image the location of their initials when the rotated image coincides with the original, and have them describe how far the traced image was rotated. The students may use fractions or degrees, depending on their level of sophistication.

Have the pairs of students explore the remaining figures on the activity sheet. Write the following questions on the board for the students to consider as they complete the activity sheet:

- Do all the figures have rotational symmetry? (no)
- How can you tell? (A figure has rotational symmetry if it can be rotated less that a full turn and the rotated image coincides with the original figure.)
- Which figures have half-turn, or 180-degree, rotational symmetry? (the cross, the parallelogram, and the letter *I*)
- Which figure has the greatest number of different angle rotations that demonstrate rotational symmetry? (the cross) Why do you think so? (The traced image coincides with the original figure three times within a full rotation. It has 90° [1/4 turn], 180° [1/2 turn], and 270° [3/4 turn] rotational symmetry.)

When the students have finished, have the pairs compare their answers and discuss the questions on the board. Draw students' attention to the figures with half-turn symmetry. Explain that they have special rotational symmetry known as *point symmetry*.

Equilateral triangle: 120° (1/3 turn) and 240° (2/3 turn) rotational symmetry

Cross: 90° (1/4 turn), 180° (1/2 turn), and 270° (3/4 turn) rotational symmetry

Parallelogram: 180° (1/2 turn) rotational symmetry

Heart: no rotational symmetry

Three-leaf clover: 120° (1/3 turn) and 240° (2/3 turn) rotational symmetry

Letter I: 180° (1/2 turn) rotational symmetry

Next, have the students gain some practical experience with rotational symmetry by investigating some familiar company logos and trademarks. Ask them to cut out company logos or trademarks from advertisements in newspapers or magazines. Have them use quarter sheets of overhead transparencies and the two blank full-circle protractors to identify which trademarks have rotational symmetry. The students can record their results in their mathematics journals and share their discoveries with classmates in a group discussion. The following is a sample journal prompt:

> Select two company logos that have different rotational symmetry. Copy or paste them in your journals. Describe the rotational symmetry you found for each logo, and use illustrations to explain why you classified the rotational symmetry as you did.

Finally, encourage the students to find other real-world objects that have rotational symmetry.

Assess

Ask the students to create for a school club or activity a logo that has rotational symmetry. Have them indicate the type of rotational symmetry in their logo design and justify their classifications with pictures or written explanations.

Extend

Ask the children to determine which company logos and trademarks have line symmetry as well as rotational symmetry. Have them consider the following:

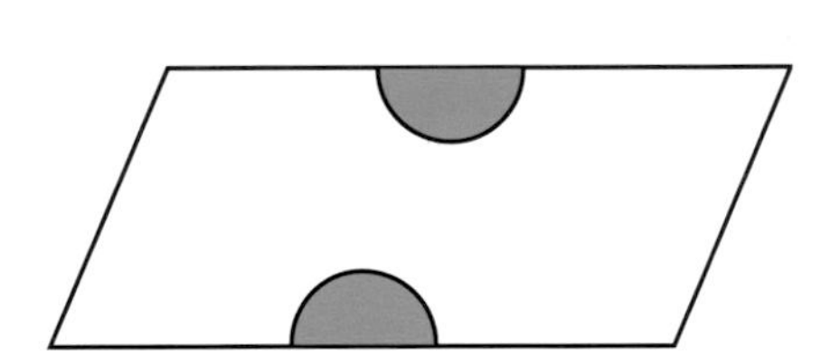

- Do all logos that have rotational symmetry also have line symmetry? Why or why not? (Not necessarily. The parallelogram logo in the margin has half-turn rotational symmetry but no line symmetry.)
- Find a company logo that has line symmetry but no rotational symmetry. (The insect logo in the margin has line symmetry but no rotational symmetry.)

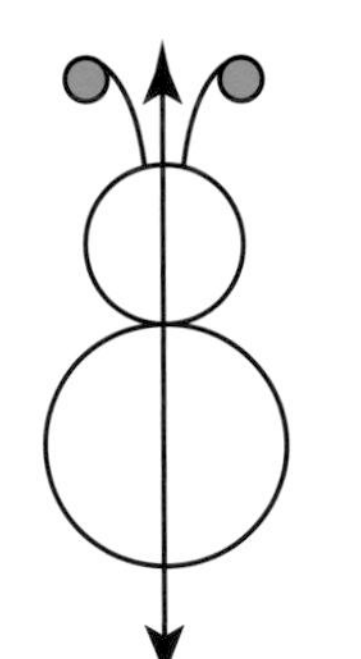

Have the students make a bulletin board displaying favorite logos that have rotational symmetry. Encourage them to investigate various national or state flags for rotational symmetry. This is one way to relate geography and mathematics. The following Web site has a variety of flags that can be downloaded from the Web site's server for this activity: http://155.187.10.12/flags/flags.html.

Tetrominoes Cover-Up

Grade 3

Goals

- Make and verify all possible arrangements of four squares, called *tetrominoes*
- Use slides, flips, and turns to completely cover a 10 × 12 grid with a variety of tetrominoes

Predict and describe the results of sliding, flipping, and turning two-dimensional shapes

Describe a motion or a series of motions that will show two shapes are congruent

pp. 120, 121

Prior Knowledge

Students should have had experience in the primary grades with turning, flipping, and sliding various geometric shapes and observing the effect of these movements on the orientations of the shapes. Students should also be familiar with such positional terms as *clockwise* and *counterclockwise*.

Materials and Equipment

- Two copies of "Tetrominoes Cover-Up Game Board" for each student
- A "Tetrominoes Spinner" for each pair of students
- One-inch square tiles
- Crayons or markers, scissors, and paper clips
- An overhead projector
- An overhead transparency of "Tetrominoes Cover-Up Game Board"

Important Geometric Terms

Translation (slide): A *translation* occurs when a figure slides to a new location without changing its orientation. *Slide* is an informal term for translation. See the examples in figure 3.5.

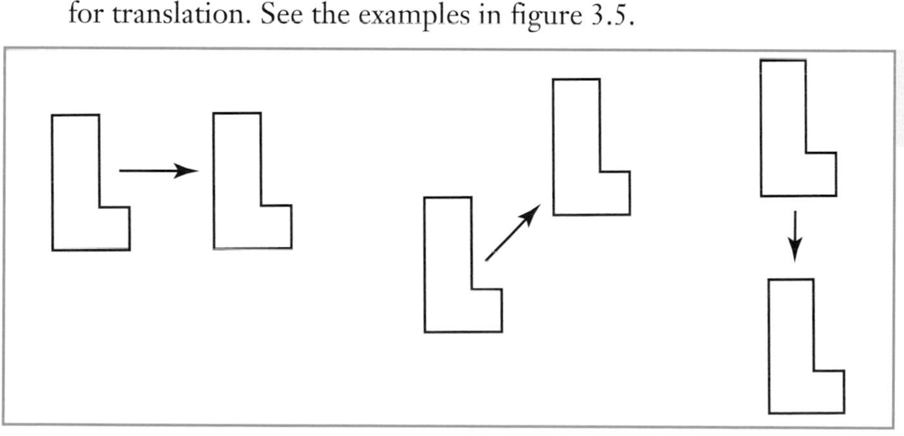

Fig. **3.5.**
Examples of translation

Reflection (flip): A *reflection* occurs when a figure is flipped so that a mirror image of the figure is created. *Flip* is an informal term for reflection. See the examples in figure 3.6.

This activity has been adapted from Clements et al. (1998).

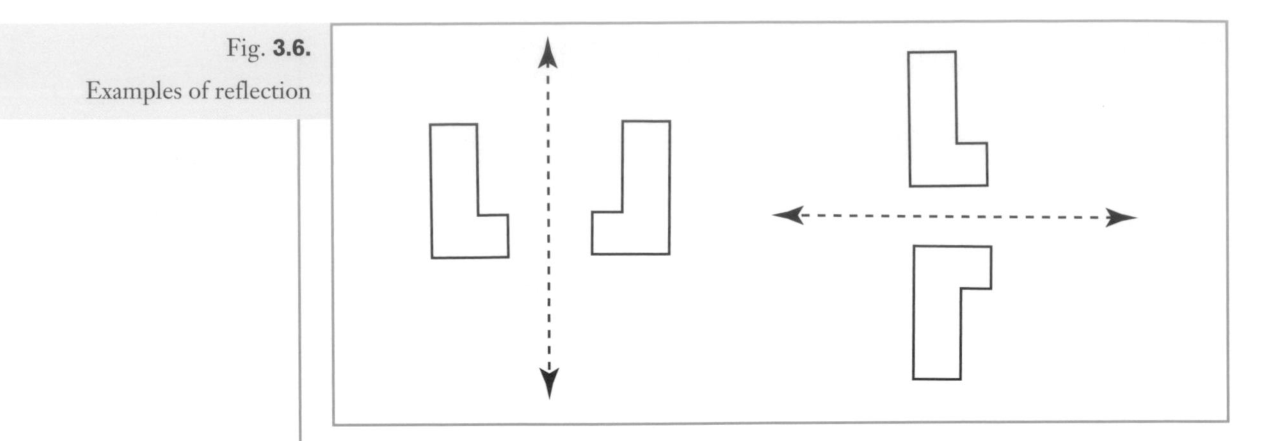

Fig. **3.6.**
Examples of reflection

Rotation (turn): A *rotation* occurs when a figure is turned about a fixed point. *Turn* is an informal term for rotation. See the example in figure 3.7.

Fig. **3.7.**
An example of rotation

A quarter turn (90°) clockwise

(It is acceptable for students at this level to use the informal names for these motions. However, it is appropriate to introduce the formal terms at this time and for the teacher to use them.)

Congruent: A term used to describe figures that are the same size and shape.

Learning Environment

The students work in pairs for all the tasks.

Activity

Engage

Have the students form pairs, and hand out square tiles to them. Challenge them to create as many different arrangements of four square tiles as possible. Demonstrate on the overhead projector the rules for arranging the tiles:

- Only complete edges may touch.
- Tiles must be laid flat; no stacking is allowed.

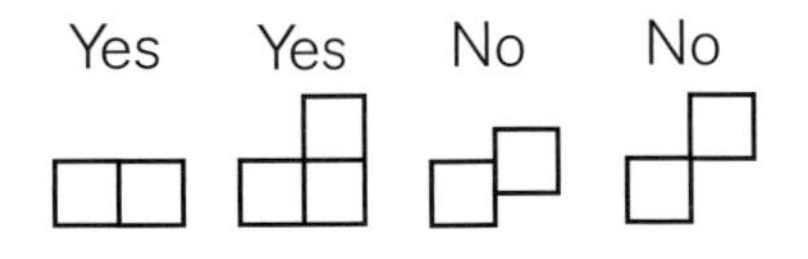

Instruct the students to record the various tile arrangements on two copies of the grid on "Tetronimoes Cover-Up Game Board," coloring each unique arrangement or shape a different color. (The game board is used as grid paper for this part of the activity.) Have the students cut out the arrangements after coloring them. Inform the students that each unique arrangement of four square tiles represents a different "tetromino." Only a certain number of unique tetromino shapes exist. Relating tetronimoes to dominoes may be helpful. A domino is a shape that is formed when two squares are put together with entire edges touching. Only one unique domino shape can be made with two squares. Encourage the students to think of names for the various tetromino shapes they discover, such as the **T** shape.

Discuss the class's discoveries, using the following questions to guide students' thinking:

- Do you have all possible tetromino shapes?
- How do you know? (There are no other possible arrangements of four tiles that make tetrominoes that are different from the five already found.)
- Are some of the tetrominoes the same?
- How can you prove it? (By turning, flipping, or sliding the tetrominoes and placing them on top of each other, we can prove they are the same or different. They are the same when they fit exactly on top of each other, proving that they are the same size and shape.)

Call on some students to demonstrate on the overhead projector which shapes are the same and which are different. They may use informal language as they discuss the motions they are using to demonstrate that two shapes are congruent. Use the terms *translation*, *reflection*, and *rotation* to paraphrase what they say, and emphasize *congruent* as a new mathematical term that means *having the same size and shape*. Demonstrate congruence as you say, "I can prove these are congruent by turning," or "I can prove these are congruent by flipping."

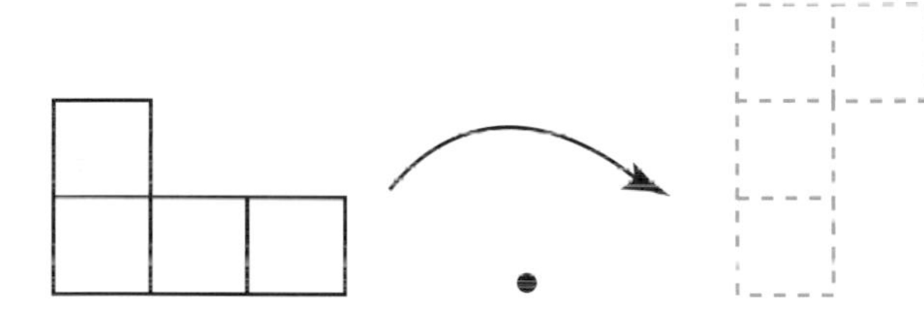

There are five unique tetrominoes:

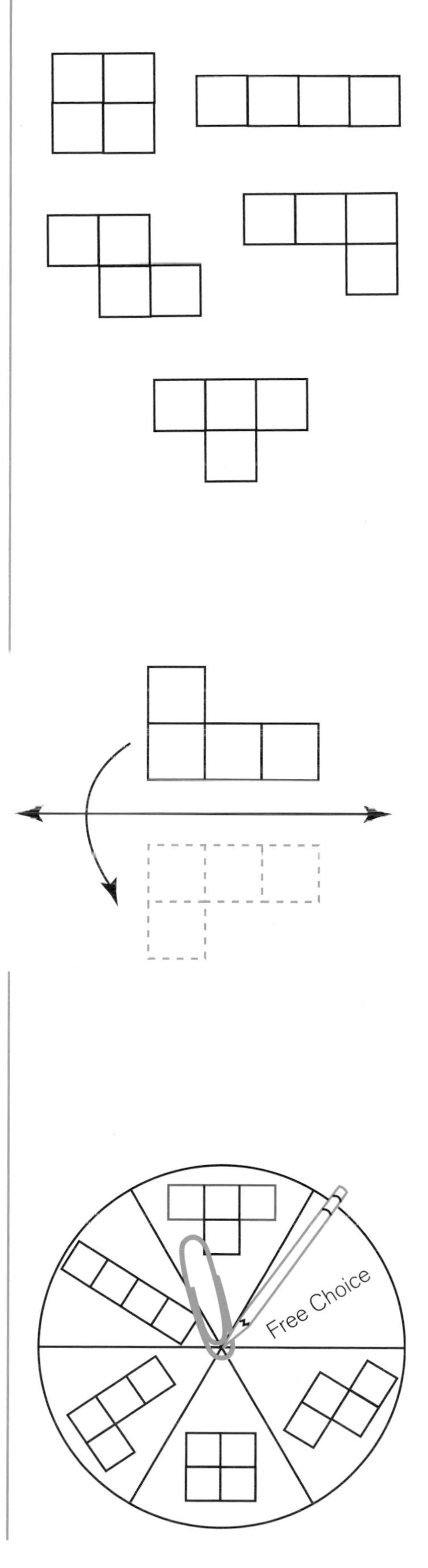

Explore

With the tetrominoes they have just made, the students play the game "tetrominoes cover-up." The object of the game is to use the tetrominoes to try to completely cover up the 10×12 grid on the game board or have the least number of uncovered squares with no overlapping. Hand out a game board to each student and a spinner and a paper clip to each pair of players. To make the spinner, place one end of a paper clip just beyond the center of the spinner. Then place the tip of a pencil or pen directly in the center of the spinner and just inside the paper clip to hold it in place when spinning. Using an index finger, push the free end of the paper clip, spinning it clockwise or counterclockwise around the pen or pencil. The section of the spinner that the paper clip lands on is the tetromino to be used for that turn. A player whose spin lands on "Free Choice" may play a piece of his or her choice.

Model on the overhead projector how to play the game. Each pair must decide who will go first. Player 1 spins to select a tetromino piece to play. Each player then takes the selected tetromino piece and holds it at the top of the game board outside the rectangular region, and then

As a variation, teachers may opt to have their students use a divider between players so they do not just copy each other's moves.

the players simultaneously move the tetromino piece downward within the playing regions. The players must place the tetrominoes on their game boards such that one side of the tetromino touches either the bottom of the game board or (after the first round) another tetromino. They may use turns, slides, and flips to place the selected tetromino so that the fewest spaces will be left uncovered on the game board. The players then color the squares that are covered by the selected tetromino.

Player 2 then takes a turn spinning to determine the tetromino to be placed on both players' game boards. Play continues until no more tetrominoes can be placed on either game board. The players must then determine their scores for the game. A player's score is the total number of squares not covered on the game board. The winner is the player with the lowest score.

A variation is to play the game using six sets of tetrominoes and only one game board. Instead of coloring squares on the game board, the students can take turns spinning to select a tetromino shape and then place the actual tetromino on the game board. When all the pieces of a certain shape have been used, the students either spin again or lose a turn. Play ends when no tetrominoes are left to play or none of the remaining shapes will fit on the playing region. The student that places the greater number of tetrominoes on the game board is the winner.

When all the students have played the game at least once, discuss with the whole class some strategies they have discovered. The following questions can be used to guide the discussion:

- Do certain shapes fit together well?
- How did you decide where to place the tetrominoes?
- Was one tetromino shape more difficult to place than the others? Why?
- What was the easiest tetromino shape to work with? Why?

These questions are open-ended, and the students' responses will vary.

Assess

Observe the students as they play the game. Try to notice whether they are planning ahead before they place the tetrominoes and what strategies they are using. In the beginning, they may use trial and error; however, after some experience with the game, they should be encouraged to visualize where to place the piece and plan their move before placing it in position. In addition, question students individually as they are playing. Some guiding questions include the following:

- Where are you going to put the piece? Why?
- What motions will you need to use to get it there?
- Where else on the game board will that piece fit?

Extend

Have pairs of students play together to fill up a game board. Using the terms *translation*, *reflection*, and *rotation*, they should take turns directing each other how to move the selected tetromino from outside the top of the game board downward onto the playing grid. They might also explore playing the game with game boards of different sizes. For example, the students might try playing on an 8×15 grid or a 6×24 grid. Have the students determine which game board was more challenging and tell why.

Where to Go Next in Instruction?

The students should continue investigating translations, reflections, and rotations. The next activity, Motion Commotion, is a good follow-up lesson.

Motion Commotion

Grades 3–4

Predict and describe the results of sliding, flipping, and turning two-dimensional shapes

Describe a motion or a series of motions that show that two shapes are congruent

p. 122

Goals

- Manipulate a figure using the following basic transformations: translations (slides), reflections (flips), and rotations (turns)
- Predict the new orientation of a figure after a specific transformation

Prior Knowledge

The students should have had some experience with turning, sliding, and flipping figures and observing the results of these actions on the figures' orientation. They should also be familiar with the positional terms *horizontal*, *vertical*, and *clockwise*.

Materials and Equipment

- A copy of the "Motion Commotion" blackline master for each student
- An overhead projector and washable overhead-transparency markers
- Overhead-transparency cutouts of the figures from the "Motion Commotion" blackline master
- Scissors, markers or crayons, and pencils
- Coffee stirrers

Learning Environment

The students work as a whole group during the "Engage" portion of the lesson and in pairs for the remainder of the lesson.

Important Geometric Terms

Translation (slide), *reflection (flip)*, *rotation (turn)*
(See the previous activity for definitions and illustrations of the terms above. See also the "Engage" and "Explore" sections of this activity.)

Activity

Engage

Cut out the notched half circle from "Motion Commotion," and place it on top of an overhead transparency on the overhead projector. Trace the figure using an overhead-transparency marker. Then slide the cutout figure to a new location on the overhead transparency. Discuss with students the word for this transformation—*translation (slide)*. Ask them to find another way or direction to slide the figure. Have volunteers demonstrate the other translations on the overhead projector (see some examples in figure 3.8). Draw students' attention to the fact that a slide does not change the orientation of a figure; the original figure just appears to have moved to a new location.

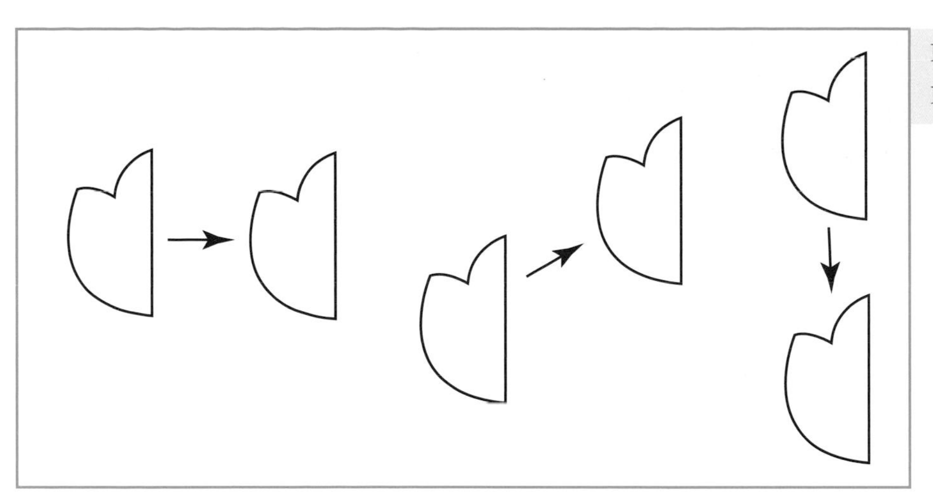

Fig. **3.8.**

Examples of translation

Next, model reflections (flips) in a similar fashion, using the same cutout figure. Use a coffee stirrer to represent a segment of the line of reflection. Draw students' attention to the mirror image that results from a reflection, and explain that the orientation of the mirror image is the reverse of the orientation of the original figure (see fig. 3.9).

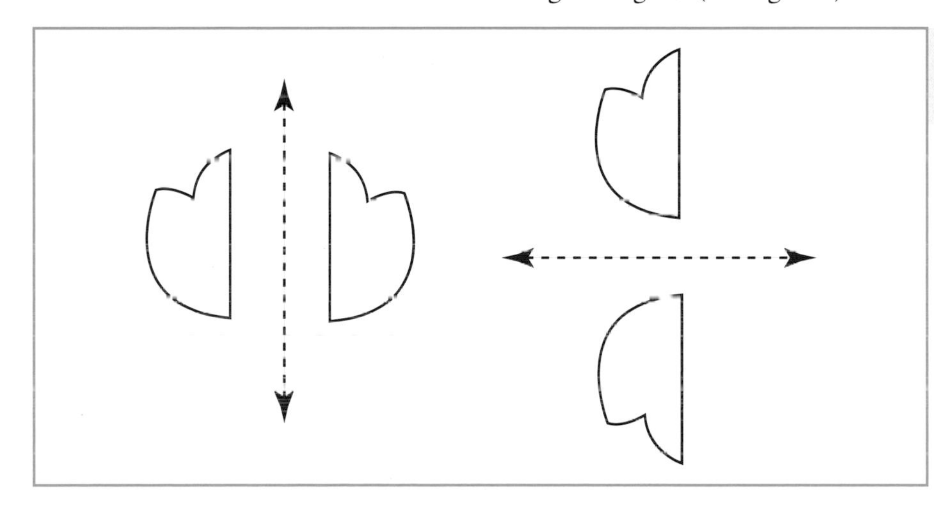

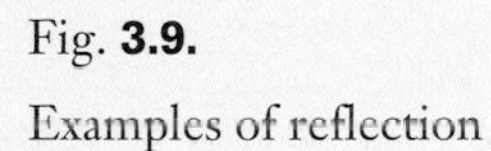

Fig. **3.9.**

Examples of reflection

Then model rotations (turns) with the same cutout. After tracing the notched half circle, hold one point fixed with the tip of a pencil and turn the cutout clockwise for a quarter turn. Draw students' attention to the fact that only one point (e.g., point O in fig. 3.10a and point P in fig. 3.10b) is held fixed while the rest of the cutout is turned.

Hand out a copy of "Motion Commotion" and a coffee stirrer to each student. Have all the students cut out the same figures from the

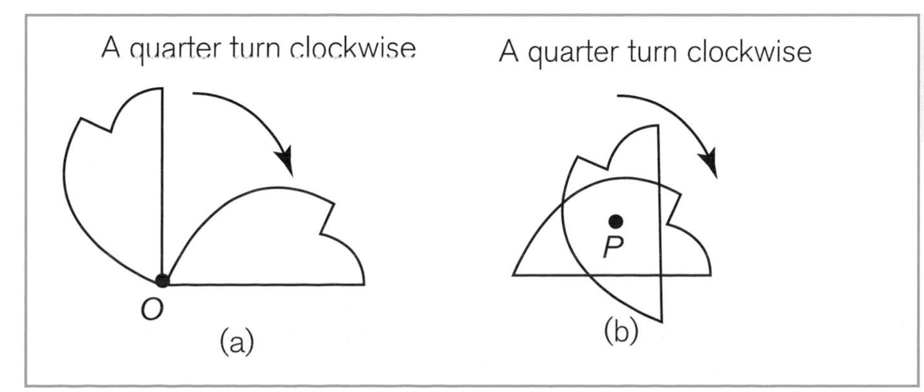

Fig. **3.10.**

Examples of rotation using different centers of rotation

This lesson incorporates only one-quarter clockwise turns. As students gain confidence with rotations, other turns should be considered. It is important that students eventually be able to recognize a variety of turns. Turns are also identified by the measure of the angle of rotation. A quarter turn is also known as a *ninety-degree turn.*

activity sheet and label the center point "A." Have the students place the figure in the upper left-hand corner of their desks in the same orientation. Play a game of "Simon says" with the class to practice translations, reflections, and rotations using the cutout figure.

Call out a variety of transformations. Be sure to describe both direction and distance when defining a translation—for example, "Simon says, 'Slide the figure three inches to the right.'" When assigning a reflection, indicate the line of reflection—for example, "Simon says, 'Flip the figure to the left over a vertical line of reflection.'" The students may use coffee stirrers to represent a segment of a line of reflection. When defining a rotation, have the students indicate the center of rotation as well as the direction and size of the angle—for example, "Simon says, 'Rotate the figure one-quarter turn clockwise about point A.'"

Explore

When the students seem comfortable with the basic transformations, pair them and have each pair make a "Motion Commotion" strip using slides, flips, and turns. Model how to create such a strip: First demonstrate how to cut out and fold the "Motion Commotion" strip. Emphasize that the students must cut along all the dotted lines so that the top half of the strip has fringelike flaps. These flaps will eventually be folded down to cover the images drawn on the strip.

Have the students select one of the figures at the bottom of "Motion Commotion" and cut it out. Ask them to trace the figure in the first (lower left-hand) box on the "Motion Commotion" strip. This box is the only one that does not have a flap above it. Then have the students place the figure in the same orientation on their desktops just below the first box on the "Motion Commotion" strip. Next, have the students perform a translation, a reflection, or a rotation on the figure and record the resulting image in the box to the right of the first one. Direct the students to fold down the flap above the second box so that it covers the new image, and have them write on the flap a description of the transformation they just performed (see fig. 3.11). Continuing with the orientation that is then on the desktop, the students should perform a new transformation, record the new image in the next box, and write a description of the second transformation on the corresponding flap. Instruct the students to continue this procedure until all five boxes have been completed.

In this activity, rotations are limited to one-quarter clockwise turns. The center of rotation has been marked on each figure. The students may slide a figure down or to the right. The distance a figure may slide is limited to the length or width of one box on the "Motion Commotion" strip. The students are permitted to flip a figure about a vertical or a horizontal line of reflection. In this activity, a vertical line is defined as one that runs parallel to the left or right edges of the "Motion Commotion" strip and a horizontal line is defined as one that runs parallel to the top and bottom edges of the strip. The students may use the coffee stirrers to represent a segment of a line of reflection.

Encourage the students to use a variety of transformations—such as two reflections, two translations, and one rotation—in their "Motion Commotion" strips. Some examples of the descriptions of transformations that students might write on the outside of the flaps include these:

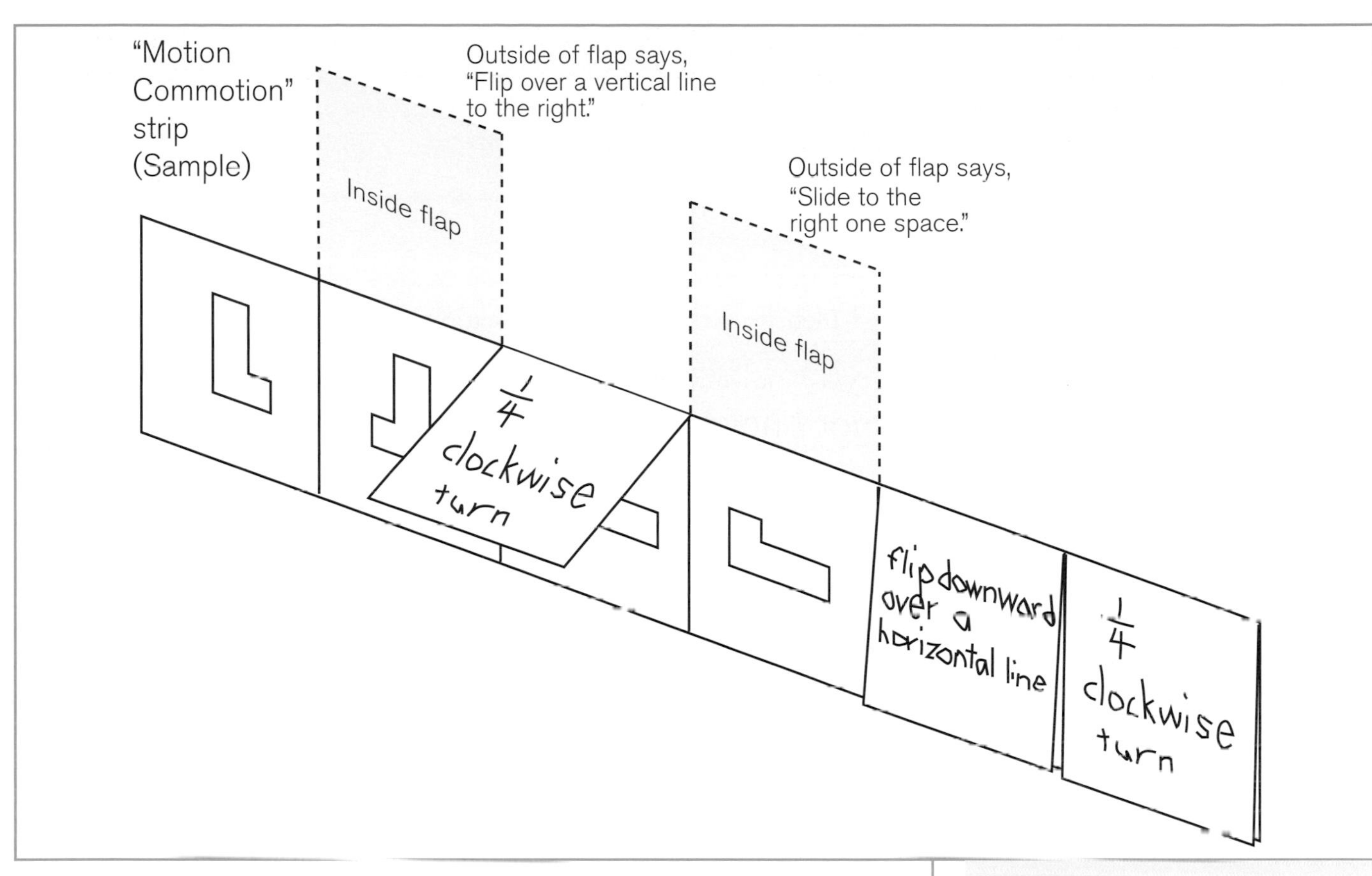

Fig. **3.11.**

A completed "Motion Commotion" strip

- Slide to the right one space
- Flip about a vertical line to the right
- 1/4 clockwise turn
- Flip downward about a horizontal line
- Slide downward to the bottom of the space

When the students have completed the "Motion Commotion" strips, have them exchange strips with their partners and predict the hidden images on their partner's strip. Have them record their predictions on the "solution strip" found on the activity sheet.

For a challenge, have the students predict—without looking under any flaps—every other image or the middle or last image.

When the students have finished working with their partners' strips, have them share their strips as a whole group and discuss any differences between the drawings and the partners' predictions.

Assess

Analyze the students' completed work to verify that the figures they drew match the descriptions on the flaps.

Extend

Have the students work with more-sophisticated rotations, such as one-third, three-fourths, and one-half turns. Encourage them to discover how half turns and reflections are related. Challenge the students to find everyday applications of these motions. Pushing a book across a desk is an example of a translation. The movement of the hands on a clock is an example of a rotation. A pair of socks hanging on a clothesline might illustrate a reflection. Have the students go on a "motion hunt" for one week and record their discoveries in their math journals.

Zany Tessellations

Grades 4–5

Predict and describe the results of sliding, flipping, and turning two-dimensional shapes

Describe a motion or series of motions that will show that two shapes are congruent

Goals

- Explore tessellations like those created by M. C. Escher
- Identify geometric transformations found in tessellations
- Create tessellations using translations (slides) and rotations (turns)

Prior Knowledge

Students should have had some experience with tessellating, or covering a surface with, congruent geometric figures such as pattern blocks. For example, students should be able to arrange several regular hexagons so that they completely cover a surface (plane) without leaving any gaps or holes and without overlaps. Students should also be familiar with geometric transformations such as translations (slides), rotations (turns), and reflections (flips).

Materials and Equipment

- Sheets of oak tag or manila folders
- 9″ × 14″ white construction paper or manila paper
- Cellophane tape, scissors, and crayons or markers
- An overhead projector
- Books of M. C. Escher art or samples of Escher's work on Web sites. (One resource is http://library.thinkquest.org/16661/escher/tessellations.1.html.)

Learning Environment

Students work as a whole class during the "Engage" section of the lesson and in pairs for the remainder of the lesson.

Important Geometric Terms

Translation (slide), rotation (turn), reflection (flip)
(See Tetrominoes Cover-Up for definitions and illustrations of these terms.)

Glide reflection (slide and flip): A transformation that involves combining a translation (slide) and a reflection (flip). This complex transformation is mentioned in the "Extend" section of the activity.

Tessellation: A covering of a surface with one or more shapes in a pattern without gaps or holes and without overlap

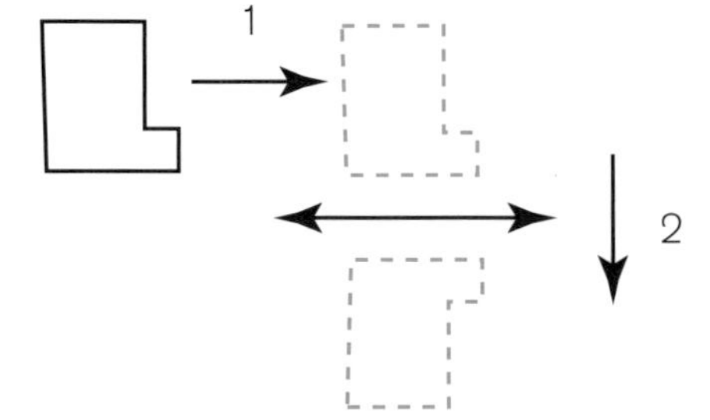

Activity

Engage

To help the students see the beauty of mathematics from an art

This activity has been adapted from Granger (2000) and Giganti and Cittadino (1990); both articles appear on the CD-ROM that accompanies this book.

perspective, introduce them to the work of M. C. Escher by displaying various examples of his work. Copies can be found in several art books or on Web sites that feature his work. See the list of additional resources for other sources. Encourage the students to use transformation terms such as *translation (slide)*, *rotation (turn)*, and *reflection (flip)* to describe the geometric shapes and repeating patterns found in Escher's tessellations. Guide the students to discover that the tessellating figures (tiles) used in his work completely cover a plane surface without leaving any gaps and without overlapping.

Explore

Tell the students that they will be drawing their own tessellations. To make interesting tessellations like Escher's, the students will design an irregular tessellating figure on a tile and then draw their own tessellations using the tile. They create these irregular figures by performing geometric transformations.

From oak tag or manila folders, cut a one-inch square for each student. Distribute the squares, and have the students lightly color one side to discourage them from accidentally flipping the pieces while they move them. Then model on the overhead projector the following steps for making an irregular tessellating figure using a translation (see fig. 3.12). This technique is sometime referred to as the "bite and push" method or the "nibble" method. It should be used with a polygon whose opposite sides are parallel and congruent. Any parallelogram will work. A square is recommended for beginners.

1. On one side of the square or other figure, draw an irregular design that goes from one corner to the adjacent corner.
2. Cut out the design on the line that was drawn. Slide the cutout to the opposite side of the figure, making sure not to flip or turn the piece.
3. Tape the cut piece in place along the straight edge of the figure.

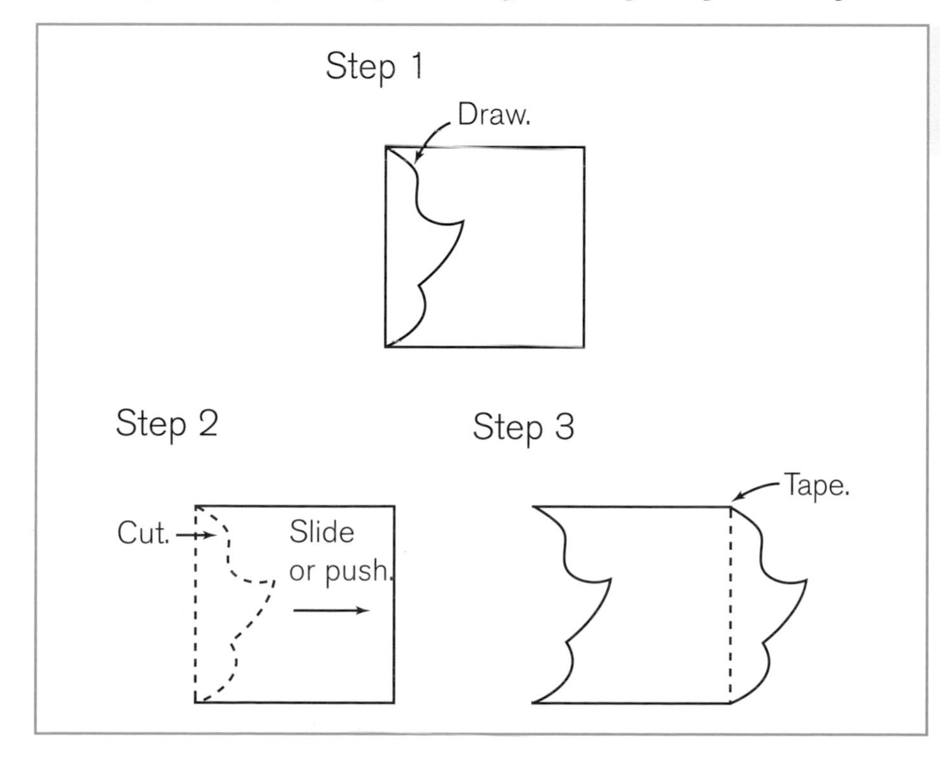

Fig. **3.12.**

The steps to making an irregular tessellating figure by translation

As the students gain confidence with the translation technique, encourage them to alter all sides of the given figure.

To create a figure that will tessellate, the new shape must maintain the same area as the original shape. There can be no extra trimming or discarding of cut pieces. Advise the students to be careful not to overlap the pieces and to make sure the straight edges and corners match before they tape them.

Allow each student to make his or her own irregular tessellating figure using the "bite and push" translation technique. Have partners help each other tape the pieces together.

When the students have finished making their irregular tessellating tiles, demonstrate on the overhead projector how to create a tessellation by repeatedly tracing the tile on paper (see fig. 3.13). Caution the students to be careful in lining up their tiles with the figures they have already traced to ensure that they don't create any gaps or overlaps. Then suggest that they color their work creatively to make an interesting artistic tessellation. Examples of students' tessellation art appear on the CD-ROM.

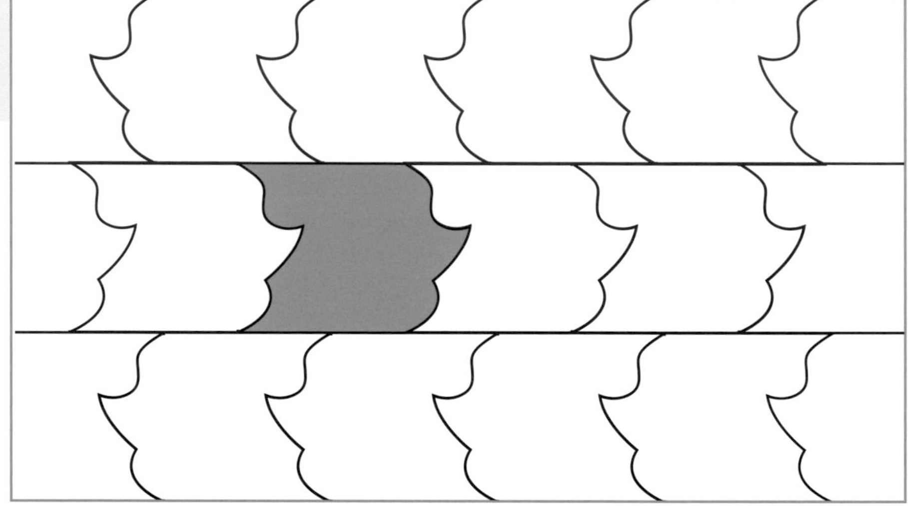

Fig. **3.13.**

A tessellation made by the bite-and-push translation method

After the students have made translation tessellations, model on the overhead projector the following rotation, or turning, technique for making irregular tessellating figures (see fig. 3.14). This technique should be limited to tessellating polygons that have adjacent sides that are congruent, such as equilateral triangles, squares, and regular hexagons. A square is recommended for beginners.

As students gain confidence with this technique, encourage them to modify or alter all sides of the square using the rotation technique.

1. Start with squares cut from oak tag or manila folders, one per student.
2. Draw a design on one side of the square from one corner to the adjacent corner.
3. Cut out the design, then rotate the cut piece using the endpoint at one corner of the cut side as the center of rotation. Rotate the piece until it lies even with an adjacent side of the square.
4. Tape the cut piece to the straight edge of the square, making sure that the pieces do not overlap and that there are no gaps.

When the students have made their tiles using the rotation technique, model on the overhead projector how to create a tessellation with a tile (see fig. 3.15). Be sure to draw their attention to the fact that

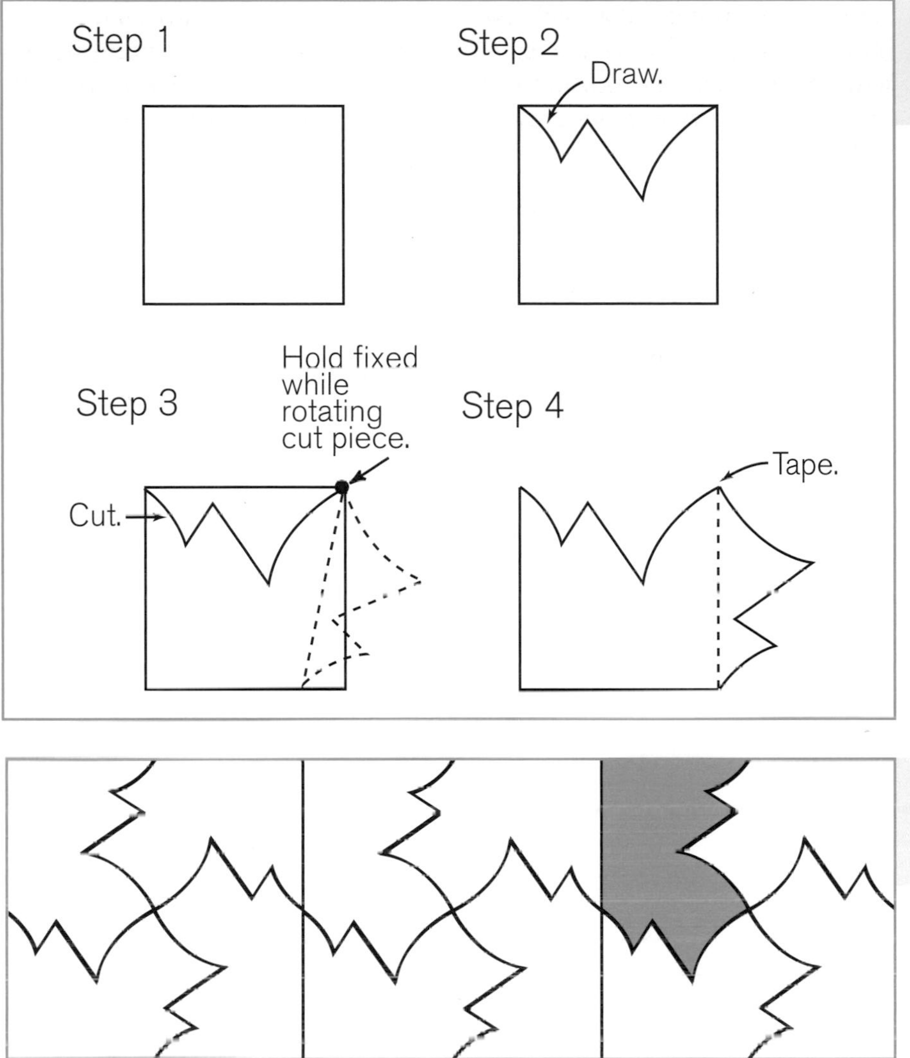

Fig. **3.14.**

The steps to making an irregular tessellating figure by rotation

Fig. **3.15.**

A tessellation made by the rotation method

this particular tessellating figure must be rotated, or turned, when they are tracing it in order to completely cover the plane surface.

As with the translation tessellations, encourage the students to use their imaginations when coloring their artwork. If the children are experiencing difficulty in identifying a design in their tiling, they may seek suggestions from their partners or other classmates.

Assess

Have the students select their favorite tessellation and attach a written description that explains the transformation technique used to create the tessellating tile. They should then identify the initial tile tracing in their artwork by marking it with an *X*. Then have the students label three other tracings in the tessellation with the letters *A*, *B*, and *C*. They should then describe on the back of their artwork the transformations used to map the initial tile tracing to the other lettered tracings. Have the students exchange papers with their partners, who then identify the transformations used to map the tile labeled *X* to locations *A*, *B*, and *C*.

Extend

Model a third technique for creating an irregular tessellating figure, the glide reflection. This technique combines a translation and a reflection. It is more complex than the translation and rotation techniques and should be introduced only after the students are comfortable with those other tessellation techniques. Again, a square is recommended for the first experience with this technique. The following steps are illustrated in figure 3.16:

1. Start with squares cut from oak tag or manila folders, one for each student.
2. Draw a design on one side of the square from one corner to the adjacent corner.
3. Cut out the design on the line that was drawn. Slide the cutout to the opposite side of the figure and flip the cutout on its vertical axis.
4. Tape the cut piece to the straight edge of the square, making sure that the pieces do not overlap and that there are no gaps.

Fig. **3.16.**

The steps to making an irregular tessellating figure by glide reflection

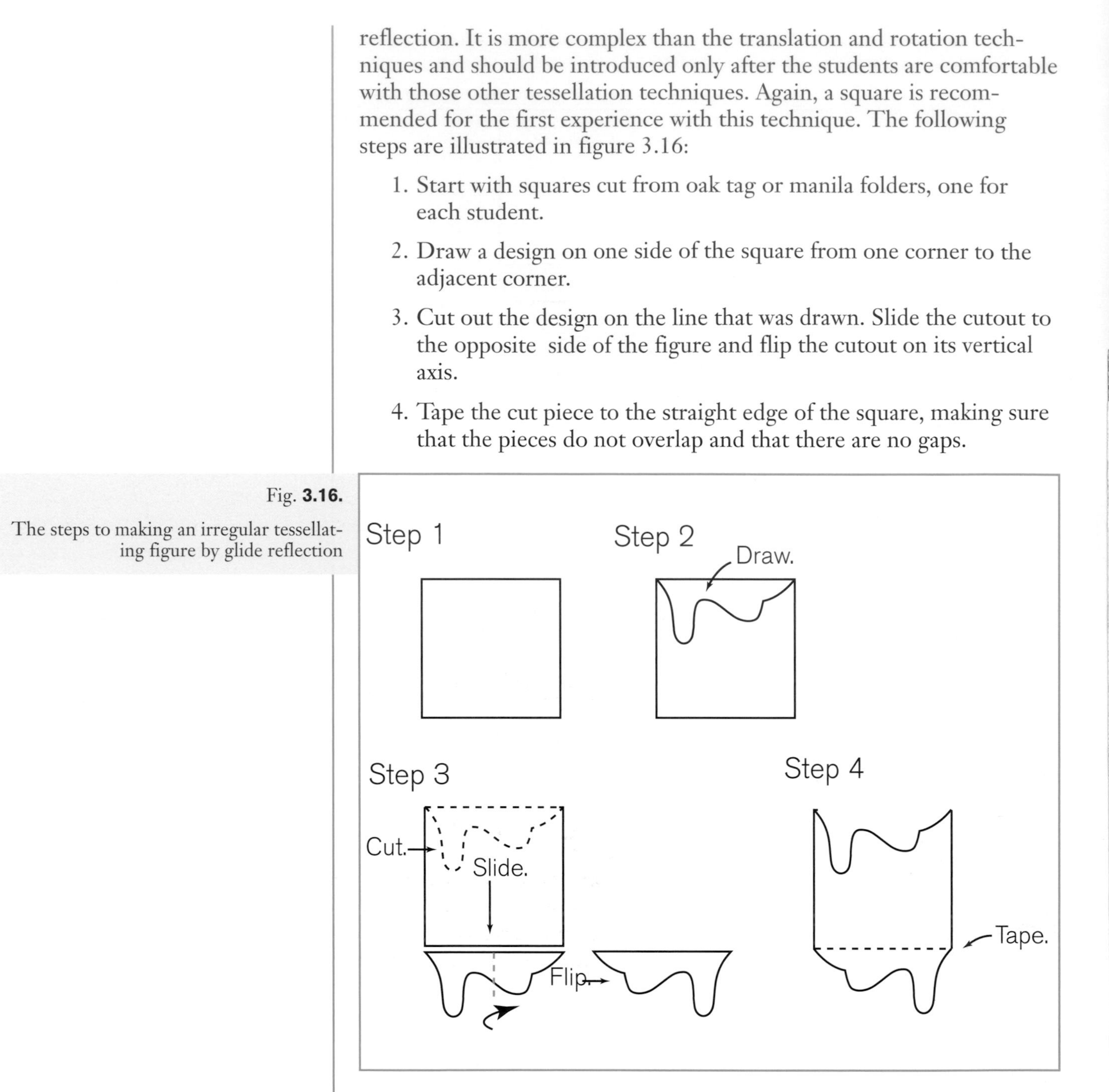

Have the students make their own glide-reflection tessellating tiles. When they have finished, model on the overhead projector how to create a tessellation using this type of figure. Be sure to draw students' attention to the fact that this particular tessellating figure must be rotated or flipped when tracing in order to tile, or cover the plane surface (see the example in fig. 3.17).

Have the students create a tessellation using their glide-reflection tiles and then color their artwork. Encourage the students to try making tessellations with more-complex polygons such as hexagons. They might also experiment with dynamic tessellating software, such as TesselMania! Deluxe (MECC 1995). After creating tessellations, the students could print their designs on T-shirt-transfer-paper using a color inkjet printer. Students could actually get to wear their artwork!

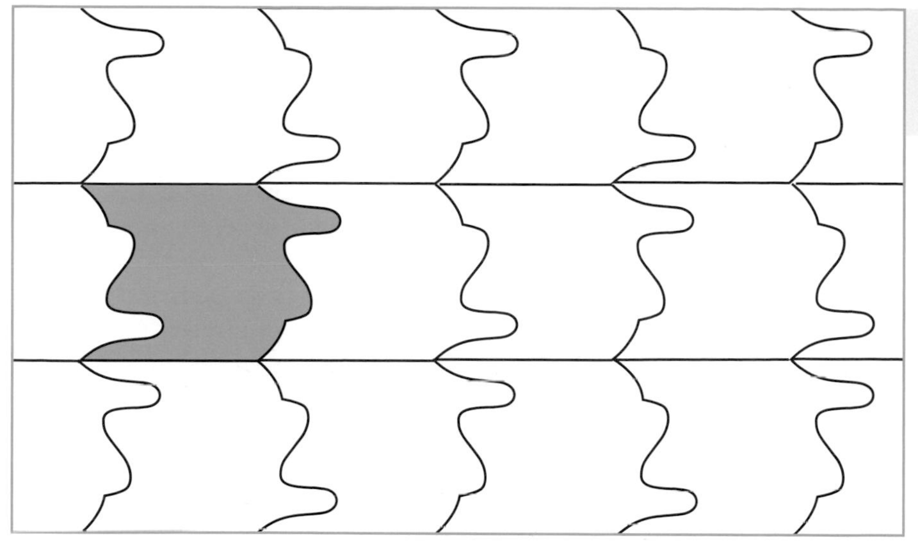

Fig. **3.17.**

A tessellation made by the glide-reflection method

Where to Go Next in Instruction?

The activities presented in this chapter give students opportunities to explore and apply transformations, as recommended in *Principles and Standards for School Mathematics* (NCTM 2000). As students in grades

3–5 investigate notions of symmetry and motion, they develop more-precise and more-sophisticated vocabulary and gain a greater awareness of how the orientation of a figure changes as it moves in space. Students discover examples of these geometric concepts in nature and in the art and architecture around them. Hence children who are exposed to these notions gain a greater appreciation of their environment.

Exploring transformations is also important for developing mental-imaging skills and spatial sense. Chapter 4, "Spatial Visualization," further develops students' intuitive sense of their environment and the objects found in it. The chapter presents a variety of interesting and engaging activities that enhance students' use of visualization, spatial reasoning, and geometric modeling to solve problems.

GRADES 3–5

NAVIGATING *through* GEOMETRY

Chapter 4
Spatial Visualization

Children develop spatial-visualization skills before they enter school. In the preschool years, children remember pathways between areas within their homes or directions to their grandmother's home. They play with blocks and puzzles, trying to fit pieces together. Teachers can take advantage of the students' early use of visual clues to build a stronger awareness and understanding of spatial visualization when they enter school.

By the end of second grade, students' geometry experiences have traditionally led them to recognize circles, squares, rectangles, and triangles. Those experiences may also have helped them develop the appropriate vocabulary to describe those shapes and some of their properties.

Changing the orientation of figures is important in developing students' mental-imaging skills as well as in helping them understand the properties of the figures they are visualizing. Students in grades 3–5 need to transform figures in their mind's eye and use the images to help construct new understandings of the figures. The activity Puzzles with Pizzazz gives them practice in these skills.

It is important to provide precise models to assist students in creating mental images. Teacher-drawn images *must* model exactly the properties of the shape the students are being asked to visualize so that the students can create accurate mental images. Accurate visualization is an important building block for further geometry experiences.

Between grades 3 and 5, spatial-visualization experiences should focus on the properties of figures, mental manipulation and the imaging of figures, relationships among figures, and two- and three-dimensional figures. Students should have the opportunity to draw, build, compare, model, and analyze two- and three-dimensional shapes. They should use a variety of materials, such as geometric shapes; puzzles and logic activities; isodot (isometric dot) paper and geodot paper; commercial manipulatives such as tangrams, geoblocks, and linking cubes; and three-dimensional packaging materials like empty boxes and cylinders. Teachers need to help students expand their knowledge of the properties of shapes to include concepts of congruence and similarity, and students should represent three-dimensional figures by drawing on isodot and geodot paper. In It's All in the Packaging and It's the View That Counts!, students use isodot paper to explore properties of shapes. Fraction Fantasy challenges students to visualize multiple divisions of a shape into congruent parts.

Students should practice mentally manipulating figures through translations, rotations, and reflections. For instance, they should think about such questions as Where would the star on the top of the cube be if you rotated the shape forward ninety degrees and then right ninety degrees? (See fig. 4.1.) They should also practice visualizing nets for three-dimensional shapes and creating three-dimensional figures from nets, as they do in Exploring Packages and in Geo City. And they should be encouraged to use real-world experiences and observations to enhance their spatial-visualization skills.

A *net* is a two-dimensional representation used to create a three-dimensional shape.

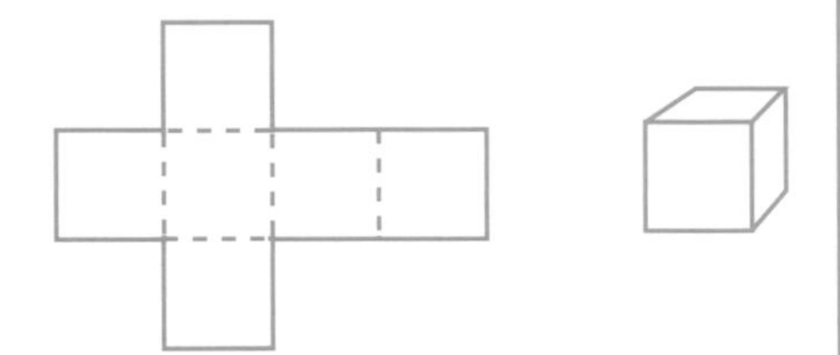

Fig. **4.1.**
A mental-manipulation exercise

Puzzles with Pizzazz

Grades 3–5

Goals

- Practice mentally manipulating shapes
- Develop strategies to solve visual logic puzzles
- Combine shapes to create different shapes

Investigate, describe, and reason about the results of subdividing, combining, and transforming shapes

Puzzles selected for students should challenge them to manipulate shapes and puzzle pieces mentally and to explore how the pieces relate to one another. Students working with jigsaw puzzles should be encouraged to use the shape of the edges of each puzzle piece to locate its place in the puzzle. Have the students use three-dimensional handheld puzzles that require them to separate the pieces and put them back together.

Create and describe mental images of objects, patterns, and paths

Prior Knowledge

The students should have had opportunities to build and solve a variety of puzzles, such as traditional picture puzzles, mazes, and hidden-picture puzzles. They should have engaged in such spatial- and visual-logic activities as tangrams, three-dimensional puzzles, and handheld puzzles that explore spatial relationships. And they should have played games like chess, checkers, and Wykersham, an old English sea captains' spatial-logic board game similar to chess and checkers.

Materials and Equipment

- A variety of spatial-visualization and logic puzzles and challenges

  Such materials might include tangrams, dissection puzzles in which shapes are reassembled or divided into other shapes, colonial cast-iron puzzles (also known as "tavern puzzles"), and three-dimensional ticktacktoe.The Tangram Challenges applet on the CD-ROM allows students to fill in the outline of a figure with tangrams and to use tangrams to form polygons.

- Puzzle problems by Martin Gardner, Charles Townsend, Henry Dudeney, or other mathematicians

  Such resources would be useful additions to a collection of spatial-visualization puzzles. The set of puzzles on the CD-ROM that accompanies this book can be used to begin a classroom collection.

- An overhead projector, overhead transparencies, and transparency marking pens
- A set of commercial tangrams or tangrams cut from the template on the "Tangrams" blackline master for each student
- An overhead transparency copy of the three puzzles on the blackline master "Puzzles"

pp. 123, 124

Important Geometric Terms

Tangram, rotate, reflect, translate, trapezoid, maze

The tangram is believed to have originated in China and was popular in the early 1800s. Some delightful literature connections for tangram activities include Grandfather Tang's Story, *by Ann Tompert (1990);* Three Pigs, One Wolf, and Seven Magic Shapes, *by Grace MacCrone (1997); and* The Warlord's Puzzles, *by Virginia Pilegard (2000).*

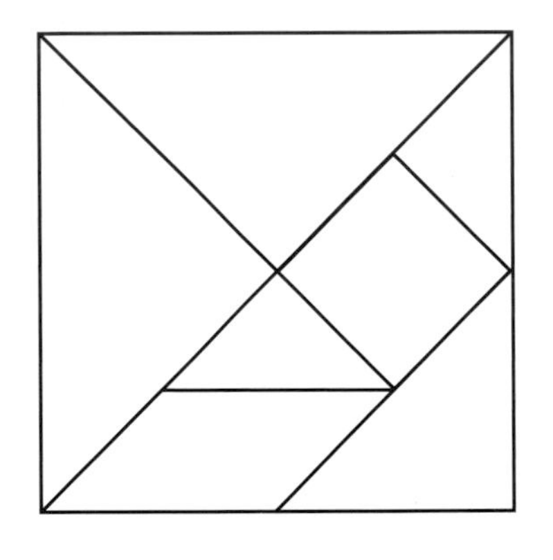

There are twelve rectangles in this shape:

The triangular figure has thirteen triangles, three trapezoids, and three parallelograms.

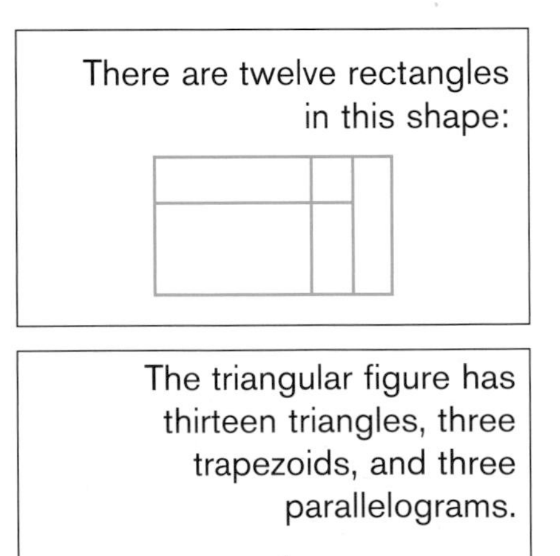

Learning Environment

The students work independently, in learning centers, or in pairs or small groups to solve a variety of puzzles and do visualization activities.

Activity

Engage

Give the students each a set of tangrams, and familiarize them with the puzzles by giving them tasks such as the following:

- Using the small tangram pieces to create one of the large tangram pieces (Ask, for example, "What pieces can you use to create the large triangle?")
- Assembling shapes to create another shape (Ask, for example, "Can you use two pieces to create a parallelogram?")

To encourage the students to manipulate a shape mentally, ask, "What tangram pieces can you use to cover a given design?" Challenge the students with questions such as these:

- How many different ways can you create a square using a set of tangrams?
- Can you use all seven tangram pieces to create a square?
- Try to create a square using one tangram piece, two tangram pieces, three tangram pieces, and so on.
- Can you create a square using four, five, and six tangram pieces?

Have the students use the tangram set to create other geometric shapes such as triangles, parallelograms, and rectangles. For instance, they could use all seven pieces to create a triangle or start with the smallest triangle and determine what pieces could be used to make a larger triangle.

Explore

Display on the overhead projector the three puzzles on the "Puzzles" blackline master. As a whole class, have the students discover how many rectangles are in the rectangle puzzle And how many triangles are in the triangle puzzle. They could also explore the number of different shapes found in the triangular figure.

To develop mental-manipulation skills, you can ask the students to mentally reconstruct a given shape into a different one. For example, they could determine without cutting out the shapes in figure 4.2 if they can be folded to create a cube. (Nets A and B can be folded into cubes.) They could also mentally rotate or manipulate figures in tasks such as these:

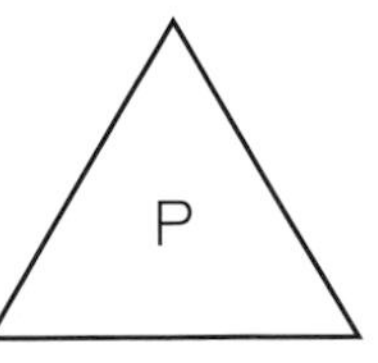

- If this triangle were rotated clockwise, how many turns would it take until the letter *P* was upright again?

  (The triangle must be turned three times, each turn through 120 degrees, before the *P* is upright again.)

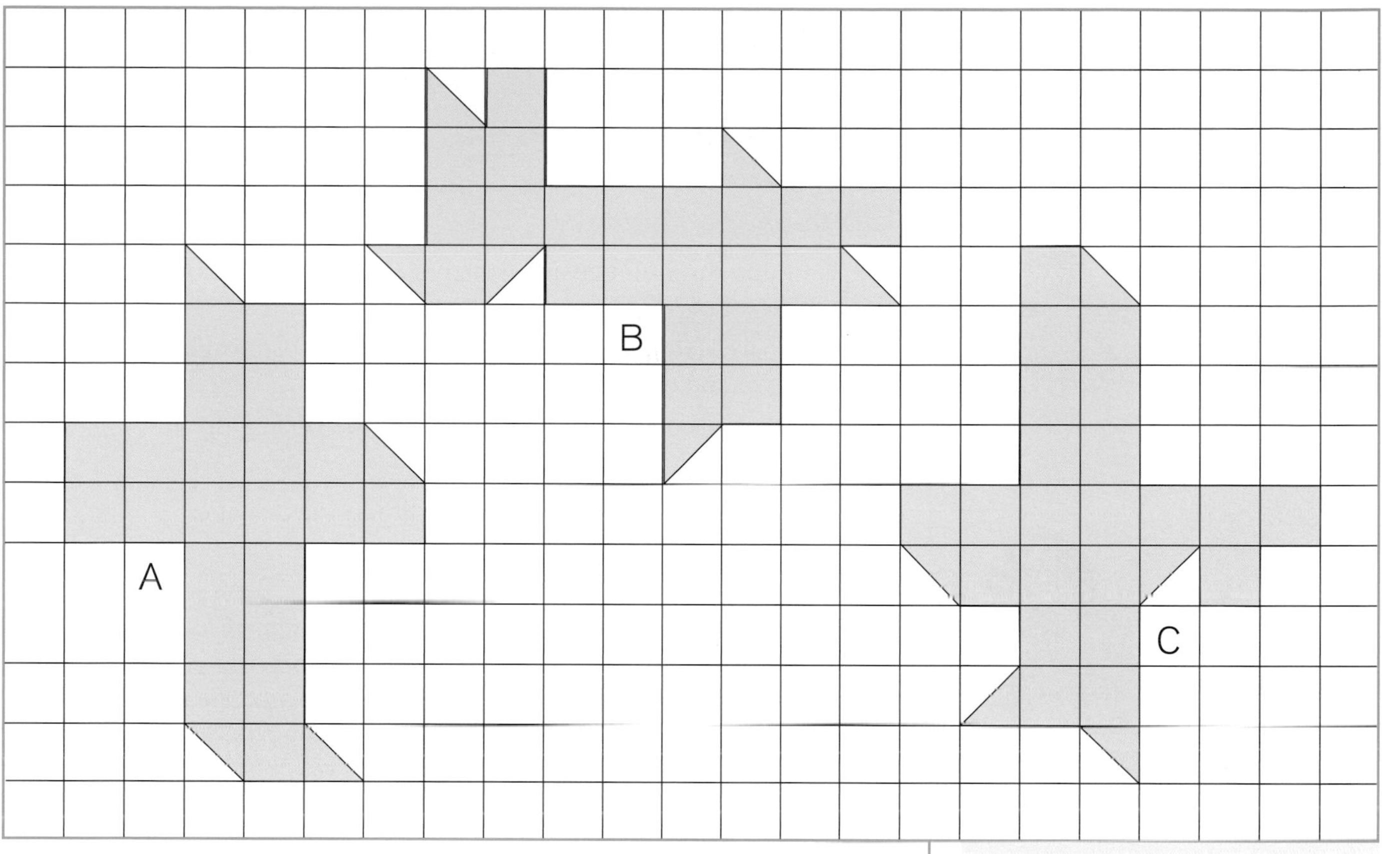

Fig. **4.2.**

Can these shapes be folded to create a cube?

- If you folded a sheet of paper lengthwise into halves and then crosswise into thirds and punched a hole in all four corners, what would the paper look like when you opened it up?

  (When unfolded, the paper would have a hole in each corner of each of the six rectanglar regions formed by folding the paper.)

- Is it possible to trace the third figure on "Puzzles" without lifting your pencil or retracing the lines?

  (One of many solutions is shown in the margin. The key is to begin the tracing at either *A* or *B*.

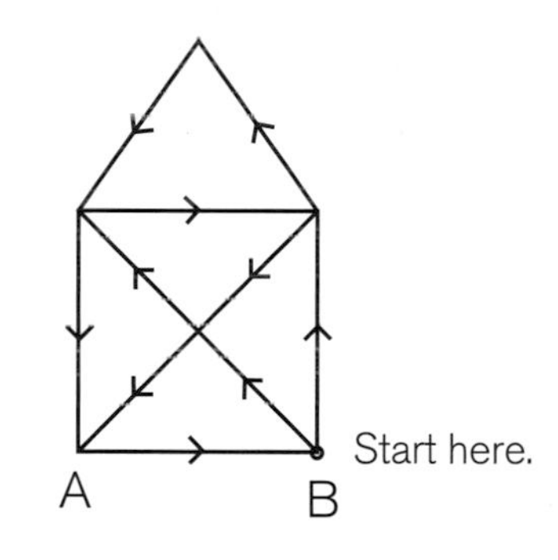

Have the students explain how they solved the puzzles or what they "saw." After you try several types of puzzles with the class, have the students solve more puzzles in pairs, small groups, or at learning centers.

Assess

Give the students a record sheet to record the different puzzles they attempted and solved. Ask them to write or explain how they solved the puzzles or to demonstrate their solutions. Have the students develop a strategy for solving a puzzle and determine if the strategy can be used to solve other puzzles.

Extend

The students could create their own puzzles and have fellow students try to solve them. The class might hold a competition for the most puzzles solved by a team of players.

A classroom organizational hint: Mount a photocopy of a puzzle on one half of the inside of a file folder, place a number on the tab of the folder, and laminate the folder. Then tape a manila envelope on the opposite side of the inside to hold puzzle pieces. The number can be used to record the puzzles the students have completed.

Exploring Packages

Grades 3–4

Identify, compare, and analyze attributes of two- and three-dimensional shapes and develop vocabulary to describe the attributes

Identify and build a two-dimensional representation of a three-dimensional object

Goals

- Examine three-dimensional shapes and create a list of their attributes
- Compare a container with a solid shape and determine how they are alike and different
- Create a net of a container by tracing the packaging
- Compare the attributes of different nets
- Explore congruent faces of three-dimensional shapes

Prior Knowledge

The students should know geometric terms, such as *face*, *edge*, and *vertex*, that are related to three-dimensional shapes and geometric solids. They should be able to recognize basic geometric shapes and identify congruent shapes. They should have had experience drawing the profiles of a three-dimensional shape.

Materials and Equipment

- An empty container (package) for each student that is a rectangular prism, such as a cereal box, a cracker package, or a milk container
- Enough geometric solids, including a cube, a cylinder, and a rectangular prism, for every student to have one
- Oak tag or construction paper
- Grid paper (Blackline masters are available for three sizes of grid paper.)

pp. 113, 114, 127

Important Geometric Terms

Face, edge, vertex, congruent, length, width, depth

Learning Environment

The students work individually in the "Engage" section of the activity and in pairs during the "Explore" portion.

Activity

Engage

Give the students geometric solids, and have them create a list of the different properties of the shapes. Show the class a cube and an empty container in the shape of a rectangular prism. Ask the students how the cube and the container are the same or different. Select a different container and have the class compare how the two containers are the same or different.

Explore

Give the students each an empty container. Have them work with a partner to compare the shapes of their containers and make a list of how the packages are similar and different. Ask them questions such as "Does your shape have any congruent faces? Can packaging be made without congruent faces?" Have the pairs record the number of congruent faces on their containers and identify them. Have them label the front, back, sides, top, and bottom of each container.

Have the students draw on grid paper a "view plan" for their package and label the views (see fig. 4.3). Then have them carefully take their boxes apart at the glue seam, lay the boxes flat on sheets of oak tag or construction paper, and trace around them to create a net. They should mark the overlapping parts of the box (the glue seams) and label the faces of the net "Side," "Front," "Back," "Bottom front," "Bottom back," "Bottom side," "Top front," "Top back," and "Top side," as appropriate (see fig. 4.4).

Ask the students to be careful when taking their boxes apart. Have spare containers available in case they tear the boxes.

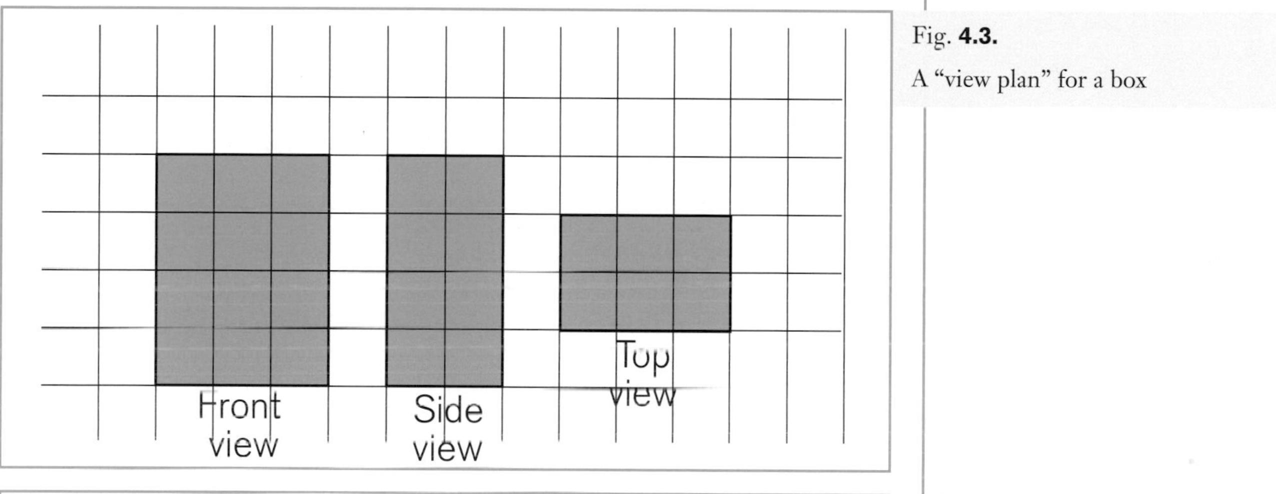

Fig. **4.3.**
A "view plan" for a box

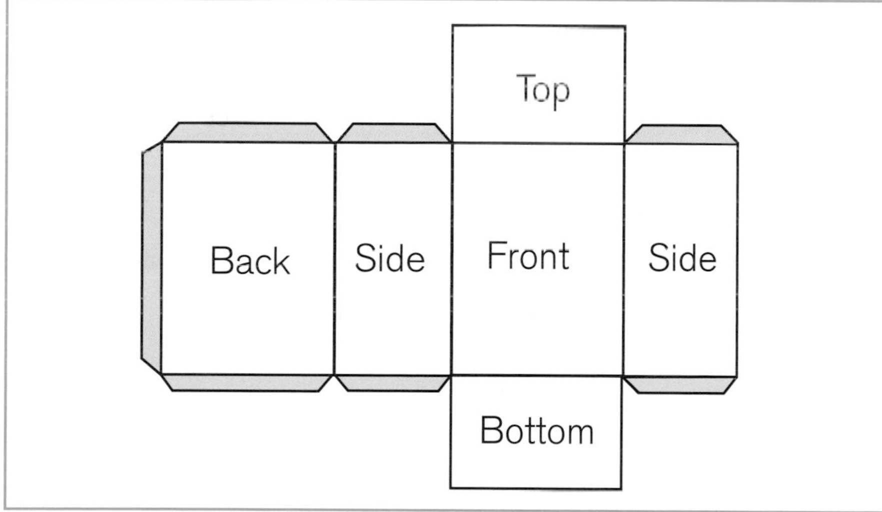

Fig. **4.4.**
A net for a box

Have the students compare the net of their box with the net of their partner's and note any similarities and differences. Ask questions such as these:

- How do the congruent faces relate to the net? (Most often, the congruent faces will not be next to each other but separated by another face. The cube is an exception to this statement.)
- How do the fold lines relate to the edges of the box? (The fold lines become the edges of the three-dimensional shape created by folding the net.)

- What shape would be built if you cut out the net and constructed a box? (The shape of the box is determined by the shape of the original packaging. For example, a cereal box is a rectangular prism.)

Have the students cut apart the original box on the fold lines, thus cutting off the glue seams, to confirm the congruence of the faces and then cut out the traced net and build a box from it.

Assess

Ask the students to identify in their mathematics journals as many properties of their three-dimensional shapes as they can. They should indicate that the shape has length, width, and depth and record the number of faces, edges, and vertices. Ask them to determine the relationship between the three-dimensional shapes and their two-dimensional nets. The students should be able to state that the two-dimensional shapes have only length and width whereas the three-dimensional shapes also have depth. The students may note that both the two-dimensional and the three-dimensional representations have faces and that two-dimensional figures have sides whereas three-dimensional shapes have edges. The students who may have described *points* or *corners* should be encouraged to use the term *vertices*.

The students can relate different materials to the type of container needed to hold them. They can explain how packaging for a product is determined and why some packages are not six-faced rectangular prisms. For example, some household room deodorizers are in packages that have five rectangular faces, with the sixth face cut away to display the product.

Extend

Collect an assortment of empty packages. Encourage the students to find containers that have unusual shapes. Sort and classify the set of empty boxes according to the shape of their nets or to the number of faces and edges.

Where to Go Next in Instruction?

The students should work with cylinders (e.g., oatmeal and potato-chip containers) and cones as well as with pyramids and prisms. The students can explore grocery stores to find candy that is packaged in a triangular prism or other products that use three-dimensional packaging that is not a rectangular prism. They should identify how shapes with edges and vertices are like and different from such shapes as spheres and cylinders, and they should compare pyramids with prisms and analyze their properties.

It's All in the Packaging

Grade 4

Goals

- Explore how three-dimensional packages are constructed
- Use geodot paper or grid paper to create front, top, and side views of packages
- Create new "packages" for a product

Build and draw geometric objects

Recognize geometric ideas and relationships and apply them to other disciplines and to problems that arise in the classroom or in everyday life

Prior Knowledge

Students should have taken packages apart to study how they are constructed. Some experience in drawing on isodot (isometric dot) paper or grid paper may help students draw three-dimensional shapes on the papers.

Materials and Equipment

- Empty packages such as cereal boxes, oatmeal containers, pretzel boxes, saltine packages, and milk cartons in quart and half-gallon sizes, one box per student
- Objects (e.g., materials brought from home, such as pasta, cereal, dog bones, or cookies, or classroom supplies, such as crayons or pattern blocks in small plastic bags) to be used as "products" for which students can design packages
- Isodot paper and either geodot paper or grid paper
- 24″ × 36″ pieces of construction paper, art-craft paper, or oak tag
- Tape, glue, and other art supplies such as markers, crayons, and colored pencils

Learning Environment

The students work in teams of three or four. They will make both team and individual products.

Important Geometric Terms

Net, dimension, three-dimensional

Activity

Engage

Tell the students they are employed by the Well-Built Packaging Company. A client of the firm wants their team to redesign the packaging for their most popular product. The new container must be appealing to the consumer and use as little shelf space as practical in the grocery store.

Explore

Have the students examine an empty box and explain in writing how the box might have been built. Make a class list of their ideas, and make sure they identify the following properties of a rectangular prism or cube: six faces, twelve edges, and eight vertices. Have the students brainstorm how a company might create packaging for a product. Guide them to understand that companies create a pattern that can be used to make the same shape over and over again. Have the students explore the different ways they can trace around six squares on grid paper to create a pattern that makes a six-sided closed figure when it is cut out and folded up. Have them record their responses on grid paper or geodot paper. (Fig. 4.5 shows the eleven different arrangements of squares that will fold up into a cube.) Explain that these arrangements are nets of the cube.

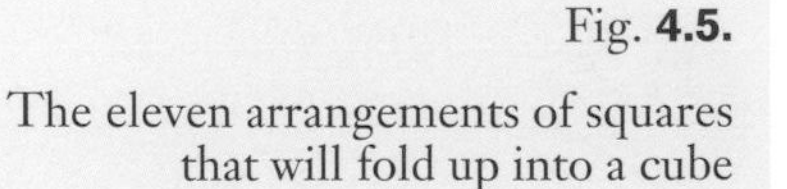

Fig. **4.5.**

The eleven arrangements of squares that will fold up into a cube

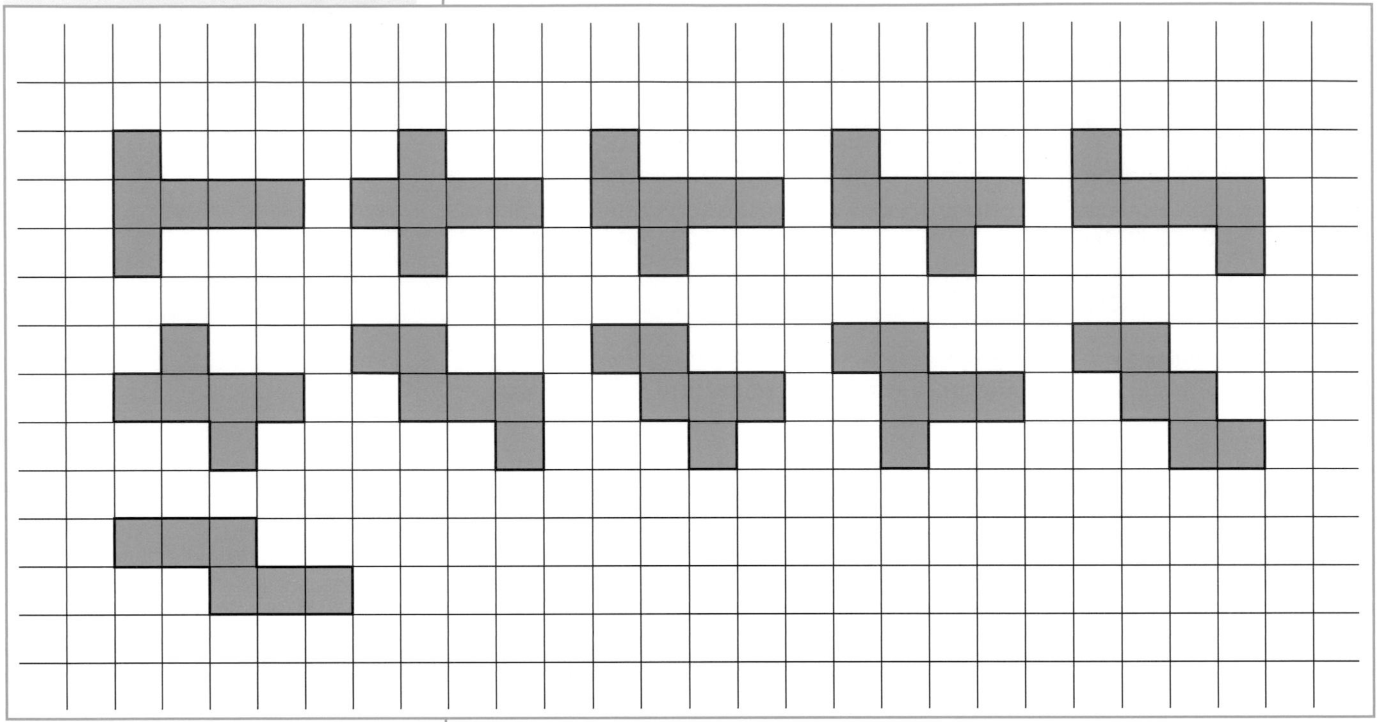

Have spare boxes on hand for students who tear their boxes.

Distribute the empty packages to the students so they can examine the different ways packages are "glued" together or tabbed to make the three-dimensional shape. Ask the students to open the package so it lies flat. Draw the net of the package by tracing around the edges. Compare the net of a cube to the net of the commercial package and determine how they are alike and different.

Once the students have explored how a package is made, give them a sample "product" and have them design packaging for it. Give each student a 24″ × 36″ sheet of oak tag or art-craft paper. Remind them that the "client" has asked the packaging company to produce a new package that is appealing and minimizes the amount of shelf space it takes up at the grocery store.

Assess

The students' work can be collected and used to assess their understanding.

Extend

Among the tasks teachers can use to extend this activity are the following:

- Develop a company logo or jingle for the "product."

- Explore the shelving of products in a grocery store to determine how many of the packages the students designed could be placed within a given space.
- Explore the careers of people who work for grocery stores or graphic designers who help design packages and market and advertise products.
- On the basis of the amount of oak tag or paper used to create the package, determine the cost of making the package the students designed.
- Draw a picture of a box on isodot paper that looks similar to the box that the students studied.

The Isometric Drawing Tool on the CD-ROM can be used to create a picture of the box.

Where to Go Next in Instruction?

The students can explore other three-dimensional figures such as pyramids, cylinders, and cones and compare the nets and properties of the figures. To begin their investigation of those shapes, they could find examples of them in art or in the world around them (e.g., the pyramids of Egypt).

It's the View That Counts!

Grade 5

Goals

Build and draw geometric objects

Identify and build a two-dimensional representation of a three-dimensional object

- Build three-dimensional objects from two-dimensional representations
- Draw representations of three-dimensional shapes on isodot paper

Prior Knowledge

The students should have drawn geometric shapes. They should be familiar with terms related to two- and three-dimensional shapes, and they should be able to describe solid shapes from different perspectives.

Materials and Equipment

pp. 125, 126, 127

- One-inch cubes, linking cubes, or geoblocks
- Isodot paper and either geodot paper or one-inch grid paper
- Rulers and pencils
- An overhead projector, overhead transparencies, and overhead-transparency markers

Learning Environment

The students work both independently and in pairs.

Important Geometric Terms

View, front, side, top, closed figure, three-dimensional shape

Activity

Engage

Give each student a set of one-inch blocks or linking cubes from which to create three-dimensional shapes. Have the students discuss what they see when they look at their constructions at eye level from the front, then from above, and finally at eye level from the side. Ask them, "How can I represent the shape so that someone else might be able to build it?" The students may respond that they would have to show the building from more that one perspective to be able to build the shape. Ask, "Who would need to draw profiles of a building and why?"

Explore

When drawing on isodot paper, do not use horizontal lines.

Give each student three blocks with which to create a three-dimensional construction, and then have the students draw on one-inch grid paper or geodot paper two-dimensional representations of the front, side, and top views of their building and label each of the views. (See the example in fig. 4.6.) The students should then share their drawing with a partner, who should try to build the shape from the view plan. Next, give each student a piece of isodot paper and one cube. Have the students draw a representation of the cube on isodot paper (see fig. 4.7). Then have them represent three joined cubes on the

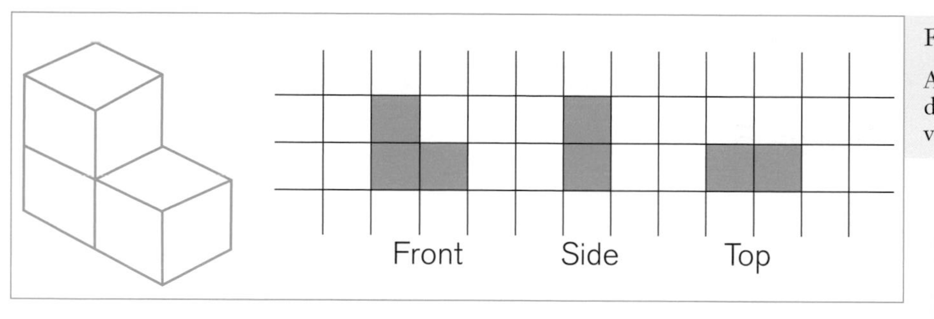

Fig. **4.6.**

A perspective drawing of a three-dimensional shape and two-dimensional views of its front, side, and top

paper. (As an alternative to drawing on isodot paper, students could use the Isometric Drawing Tool on the CD-ROM.) Once they have developed facility with the isodot paper, have them make three-dimensional structures from both view plans and isodot-paper drawings. Also have the students compare and record the similarities and differences between the isodot-paper representation and the grid-paper views. (The grid-paper drawings show the faces of the three-dimensional shapes from different views but not the depth of the configuration. All three sides must be seen to determine the exact structure from view plans. In isodot-paper representations, however, fewer views may need to be seen in order to represent all the dimensions of a structure.)

Fig. **4.7.**

An isodot-paper representation of a cube

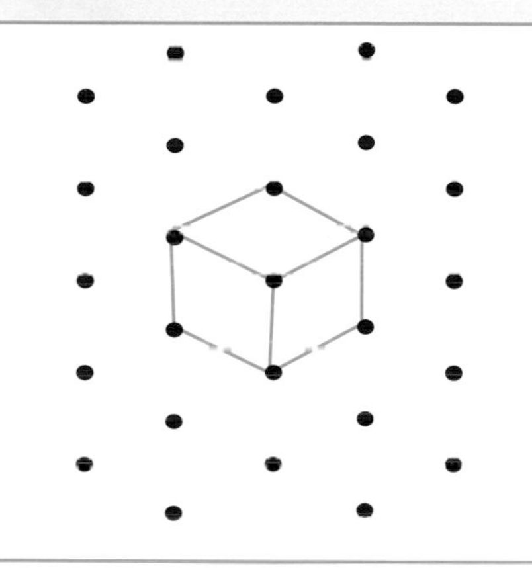

Assess

Use the students' view plans and isodot-paper drawings to assess their learning. Check for parallel lines in both representations. Make sure that the students have no horizontal lines in their isodot representations and that the cubes accurately represent the height, width, and depth of the actual structure.

Extend

Once the students understand and can create appropriate representations of three-dimensional shapes on isodot paper, have them increase the number of blocks and the complexity of the structures (see some examples of students' work in fig. 4.8). The students can also draw their names or block letters on the isodot paper.

Fig. **4.8.**

Examples of students' drawings of structures on isodot paper

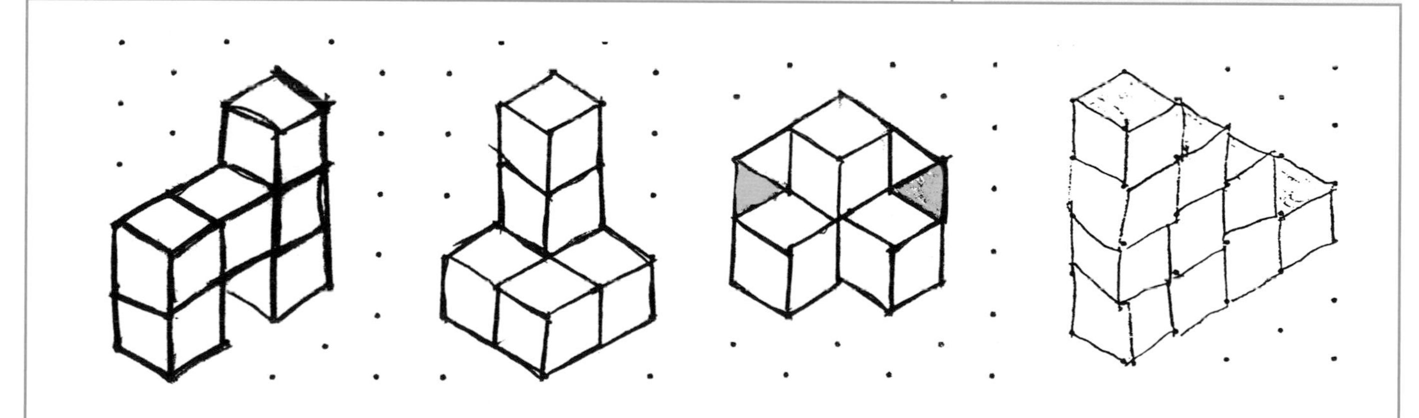

Fraction Fantasy

Grade 5

Investigate, describe, and reason about the results of subdividing, combining, and transforming shapes

Explore congruence

Predict and describe the results of sliding, flipping, and turning two-dimensional shapes

Goals

- Create and then cut from a six-inch square congruent representations of halves, thirds, fourths, fifths, sixths, eighths, tenths, and twelfths. The students must be able to re-create the original square using the fractional representations they cut.
- For each of the fractions, create multiple representations that are not congruent to any previously cut models
- Identify rotations, reflections, and translations found in the fraction models that have been created

Prior Knowledge

The students should have an understanding of fractions, symmetry, and congruence.

Materials and Equipment

- A large supply of six-inch squares cut from construction paper
- Scissors, rulers, and pencils
- An overhead projector, overhead transparencies, and overhead-transparency markers

Learning Environment

The students work individually to explore the concepts. After they have completed the task for a given fraction, the students' work can be shared.

Important Geometric Terms

Fraction, rotation, reflection, translation, symmetry, congruence

Activity

Engage

Give each student one construction-paper square. Ask the students to cut a model for the fraction one-half from the square in such a way that two congruent pieces result. Give the students another square, and ask them to create a different model for one-half that is not congruent to the models they created in their first attempt. Have them continue to cut out as many additional representations for one-half as they can that are not congruent to any they have previously created. Because this activity requires students and teachers to think differently, it is challenging. Students easily create triangular halves by cutting the square on the diagonal and also rectangles by bisecting the square horizontally or vertically. After creating the obvious models, students often reach an impasse and become frustrated because they cannot think of another model for one-half. They should be allowed to feel some "disequilibrium." Encourage them to try other strategies, such as folding the

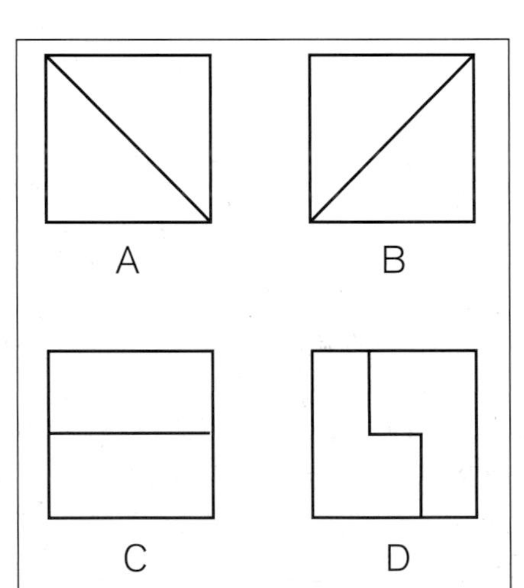

A and B, when cut, create congruent triangular shapes and therefore represent only one model for the fraction one-half. C and D, however, are two different additional models for the fraction and therefore meet the conditions of the task.

paper and using the fold lines to create a model or drawing a grid on the square and using the grid lines to create a new model.

Explore

Once the students have created several models for halves, have them move on to representing thirds, fourths, sixths, eights, tenths, and twelfths. As they create each of the models, have them keep a journal and respond to questions such as the following:

- Was it easy to create multiple models for fourths? Sixths? Eighths? Tenths? Twelfths? Why or why not? (Usually, models for fractions with even denominators are easier to create than models for fractions with odd denominators.)
- How can you prove that two fractional pieces are equal parts of the same whole? (If the parts are congruent, use transformations to show that if rotated, reflected, or translated, the parts can be made to coincide.)

Assess

The models students produce for each fraction and their journal responses can serve as indicators of their skill as they progress through the activity.

Extend

Have the students write in their mathematics journals what they think would happen if they used a different shape, such as a circle or a triangle, instead of a square to create fraction models. Then have them try to make with the selected shape the models that were challenging to create with the square. Ask, "Did your experience match your prediction? Were any of the models easier to create with the new shape? Why?"

Geo City

Grades 3–5

This extended project provides a range of activities that incorporate the geometric concepts of shape, location and coordinate points, transformations, and spatial visualization. It illustrates how to effectively integrate mathematics instruction with other content areas. The project will take from several class periods to several weeks to complete. As you read through the suggested activities, select those that are appropriate for the grade level of your students.

Goals

- Explore the relationships between two- and three-dimensional shapes (Shapes)
- Explore the effects of rotating, reflecting, and translating three-dimensional shapes (Transforming shapes)
- Create a front view of a city block on graph paper or geodot paper (Spatial relationships and spatial visualization)
- Apply mapping skills and strategies to the Geo City constructed by the class (Location and points)
- Integrate real-world knowledge of a city block in creating a three-dimensional model of a city from packaging materials and then use art skills to complete the design of the city block (Connections)
- Use the Internet to explore types of buildings (Applications to architecture)

Prior Knowledge

Students should have a basic knowledge of geometric shapes, and they should have had some experience in exploring three-dimensional shapes.

Materials and Equipment

Make available two or three boxes for each of the students because they may use several boxes to create one building.

- Empty packages, such as cereal boxes, oatmeal containers, pretzel boxes, saltine-cracker packages, and milk cartons
- Art or craft paper on a roll, large sheets of oak tag or construction paper, or rolls of brown wrapping paper to cover the boxes and create the buildings
- Tape, glue, markers, crayons, and colored pencils
- A designated area for laying out the Geo City
- Empty stove and refrigerator boxes or another source of large sheets of corrugated board
- One copy each of the "Geodot Paper" or "Quarter-Inch Grid Paper" blackline masters for each student
- One copy of the "Building Permit Application" blackline master for each team of students.

pp. 126, 114, 128

Learning Environment

Students work in teams of three or four to complete a set of buildings and such public facilities as streets and bridges that represent one block of the Geo City. All the teams join their blocks to create the class Geo City. The activity is exploratory, and all students should be able to participate.

Important Geometric Terms

Translation, front view, side view, top view, rotation, face, edge, vertex, corner, side, reflection, closed figure

Activity

The Geo City project evolves in phases. In phase 1, the students investigate cities and the types of buildings found in a city. They determine which services and related buildings are essential to a community and which are nonessential.

Phase 2 is the planning phase. The students assume the role of junior city planners whose job is to create a city in which they would like to live. The class is divided into teams, each of which is responsible for designing one block of the Geo City. Each team must submit a block plan to the classroom teacher, the chief planner for the Geo City. All teams must have the teacher's approval for their block plan before any buildings are constructed. One structure on each block must be a building the class has determined is essential for the community, such as a firehouse or a hospital. Once its plan has been approved, a team is given a large sheet of corrugated board on which to build its city block.

Teachers should emphasize the importance of planning the entire "city," which will make the job of the "chief planner" easier. To guide the students as they plan their block, each team could be assigned a different type of area. One team, for instance, might build a residential neighborhood, which could include apartments and row houses, along with schools, consumer services (gas stations, grocery stores, doctors' ofices), and perhaps a fire house, a small neighborhood restaurant, and the like. Another team might be assigned a downtown area with high-rise office buildings having street-level shops and restaurants, offices for major city services (city hall, court house, etc.), hotels, and so forth. When planning their particular buildings, the individual team members should be mindful of the kinds of buildings that are appropriate for their assigned section of the city, and the class can avoid the duplication or absence of important features of the city.

In phase 3, the students work as a team to build the essential building. Once that building has been inspected and approved by the chief planner (the teacher), all the students may create their own individual building for their team's city block.

In phase 4, all the city blocks are gathered in a large area, such as a cafeteria or a gymnasium, or on a macadam surface to create the Geo City. Wide paths representing city streets separate the city blocks. The street signs and block plans are then used to create a map of the Geo City, which the students use to navigate through the Geo City streets to locate different buildings. In the map-reading activity, the students can find the shortest route to a place or draw on the city map as many different ways as can be found to get to a specific building.

Engage

Students from the suburbs and rural areas might need an explanation of the term city block. *Teachers can log onto www.montvilleschools.org/mohegan/buildings to explore city blocks and buildings.*

Invite a local planning official to visit the classroom and talk to the students about the process of planning a building and a city and about zoning requirements. If possible, take a walk around a city block to explore the types of buildings and businesses and to observe the shapes of buildings. Hold a brainstorming session to create a list of the kinds of buildings the students might want in the Geo City. Discuss the types of services and related buildings that all cities need, such as firehouses, hospitals, schools, and police stations, and denote the essential buildings on the list.

Explore

In approving or denying applications, the chief city planner (teacher) should bear in mind that a viable city must have a healthy balance of types of buildings and businesses. Unnecessary repetitions should be avoided.

Divide the students into teams of junior city planners whose job is to design and build a block for the Geo City. Explain that the chief city planner (the teacher) must approve all building plans before construction can start, so the teams must submit applications with construction plans for their blocks before they can begin building. The plan should include one essential building and one building that each team member would like to build for the team's city block. Team members must make a "blueprint" of their block to submit to the chief city planner for approval.

Have the teams decide what buildings they would like to make with their empty containers. Allow the students to experiment with the packages to determine how the materials could be used to create many different types of buildings. Guide the discussion to consider buildings in a wide range of shapes, such as a department store, a high-rise apartment building, a bank, a museum, or a fast-food restaurant. Begin the exploration by modeling how a cereal box can be used to create different types of buildings. In the normal orientation, for instance, the front of a cereal box can be used for an office building. In the upright position, the side of the cereal box might be seen as a high-rise apartment building, whereas the same side might be viewed as an automobile showroom by laying it on its front or back. Turned so that the bottom is viewed as the front, the box can be used for a fast-food restaurant. (See the examples in fig. 4.9.)

Encourage the students to combine containers to produce more-realistic buildings. A cylindrical container, for example, can be joined to

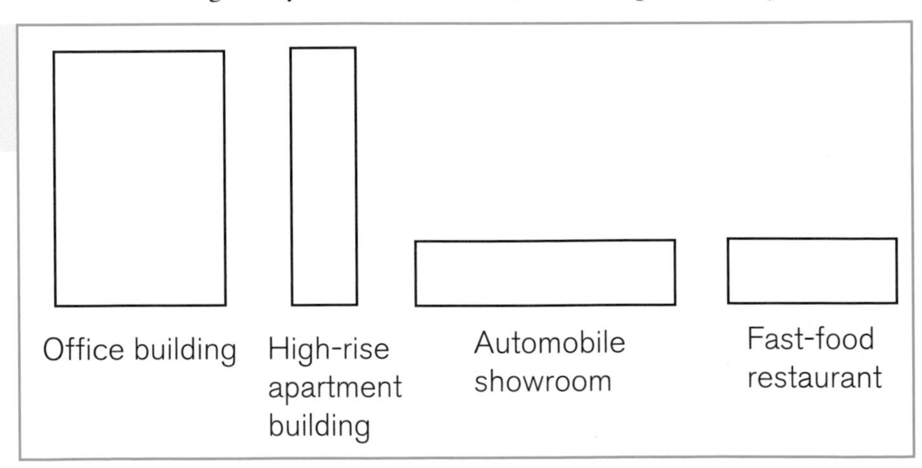

Fig. **4.9.**
Various city buildings that can be created from a cereal box

the corner of a cereal box to make a library or bank, or a cereal box can be joined with another box of the same size to produce a large office building.

The students should begin their block design by working together on plans for the essential building. From the list of city buildings, each team can choose one structure that was denoted necessary and then select the box or boxes needed to create the "blueprints" (nets of packages) for the building. Ask questions such as "What will you do with your boxes to create your building?" Encourage the students to use phrases like *I'll rotate* (or *turn*) *the box* or to talk about the similarities between the shape of the box they selected and the buildings they saw on the Web site or on the tour of the city.

Model the process of designing the blueprint for the students:

- Label all the faces of a box 1–6, as shown in figure 4.10a. Write the number 1 with a "foot" so that the students can always see how face 1 is oriented.
- Place the box, with face 1 down, on a piece of paper that is large enough to accommodate the net. Draw the outline of face 1, and label the outline "1" (see fig. 4.10b).
- Rotate the box so that face 2 is down, then draw its outline and label it "2" (see fig. 4.10c).
- Continue outlining faces 3–5.
- Ask the students how to rotate the box so that you can outline the sixth face.

A complete net is illustrated in figure 4.10d.

For buildings that are designed with multiple boxes, make a separate net for each box.

Distribute a copy of the "Building Permit Application" to each team. The teams should attach their "blueprints" to the application and submit the papers to the chief city planner, who can approve applications or request revisions to the plans. After all a team's plans have been fully approved, the students can proceed with actually building the team's block and the Geo City:

- Instruct the students to cut out the nets and fold them on the lines to get a sharp edge for the box. They should then glue the nets to the boxes, and after the glue has dried, they can use markers, crayons, and pencils to decorate and identify the "business" or other structure.
- Give each team a large sheet of corrugated board. The buildings should be attached to the corrugated board, using a low-melt-temperature glue gun if possible. Alternatively, white glue will work, but allow two days' drying time and make sure that the students have applied enough glue to hold the buildings to the board.
- Help the students lay out the city blocks of all the teams in a large space such as the gymnasium or cafeteria. Create city streets by making wide paths between the blocks. Show the students the "Sample Geo City Map" as a sample layout of the streets and buildings in a Geo City.

Use the Geo City as the basis for further geometric investigations such as the following:

- Develop path activities for the students, such as determining the shortest route between two locations in the Geo City.

Fig. **4.10.**
Creating a net for a box

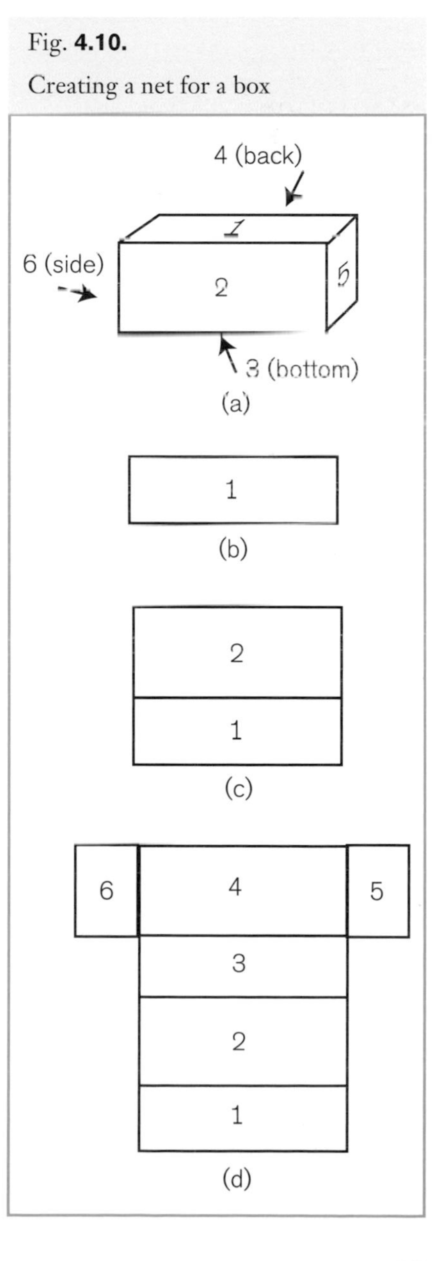

- Have the students go on a "geo search" to identify the two- and three-dimensional shapes found in the Geo City or in the buildings found on www.montvilleschools.org/mohegan/buildings. Make a class list of all the shapes found in the Geo City.
- Have the teams list an example of a translation, a rotation, and a reflection in their city blocks.
- Have the students create maps of the Geo City.

Extend

Once the teams have completed their blocks, have each student create a front view of his or her building on graph or geodot paper. (The blackline masters "Geodot Paper" and "Quarter-Inch Grid Paper" can be used for this purpose.) Then have the teams join the individual front views of the buildings to create front views of the city block. The students might use one-inch wooden cubes lined up across the front of the building to determine the number of squares to be shaded to represent the building. The blocks can be stacked to determine the height of the representation.

Assess

Observe the ease with which the students manipulate geometric shapes and the cartons to create the desired buildings. Examine the students' net designs for accuracy, and determine how well the nets fit the cardboard packages the students used for their buildings. Compare the teams' block plans with the final designs: How careful was the teams' planning? Did they have to make many adjustments? How efficiently did they solve any problems that arose? Ask the students to describe the two- and three-dimensional shapes they find in the Geo City. Have each student keep a record of his or her work in journal entries that describe—

- the process used to select geometric shapes;
- the process of constructing the buildings;
- each of the blocks of the Geo City.

Summary

In chapter 4, "Spatial Visualization," students have been prepared to manipulate shapes and figures mentally by viewing shapes and objects from different orientations. They should have begun to understand what happens when shapes are divided into regions and then reassembled to create a different shape or region. Students should have made predictions about how a shape would transform, or change, through a given motion. They should have identified and compared two- and three-dimensional shapes as well as created nets to build three-dimensional shapes. They should have learned the geometric terms related to three-dimensional shapes and represented three-dimensional figures graphically. In studying packages, the students should have begun to develop an understanding of congruence, and in the fraction activity, transformations were used to create fraction models. Through the activities in this chapter, students should have developed a better understanding of the importance of geometry in the world around them.

GRADES 3–5

NAVIGATING *through* GEOMETRY

Looking Back and Looking Ahead

Children come to school with spatial sense and preconceived notions of geometry. For example, they might connect a ball with which they are playing to a circle or they might order boxes of different sizes from smallest to largest. In the primary grades, they build on their intuitive notions and experiences by playing with, observing, and exploring a variety of shapes and structures in the school environment. In grades 3–5, students need to expand those explorations further. Their reasoning ability helps them develop mathematical arguments that permit investigations of increasing complexity. They move beyond describing shapes to classifying shapes on the basis of their properties and then to relating shapes on the basis of the similarities and differences in their properties.

The study of geometry in grades 3–5 requires a lot of *doing* as well as thinking (NCTM 2000). Sorting, building, drawing, manipulating, and measuring are vital to a deep and lasting understanding. Students' geometric explorations build the foundation for conjectures that can be tested as reasoning abilities develop. In structuring geometric experiences for children, Dina van Hiele-Geldof and Pierre M. van Hiele conclude that children pass through levels, or stages, of understanding. The van Hieles assert that the "lowest" level of understanding geometry is the visual level, in which figures are judged by their appearances: "I know it is a triangle because it looks like one."

At the next level, the descriptive level, students begin to describe the properties of the figure but the properties may not be logically ordered. For example, children may describe a triangle as a closed figure with three sides and three angles but fail to recognize congruent angles in the equilateral triangle.

The CD-ROM includes two articles that explore how children learn geometry. Pierre van Hiele (1999) discusses developing geometric thinking through activities that begin with play. Deborah Schifter (1999) explains how to study students' thinking through teachers' written observations of classroom episodes. A third article (Battista 1999) gives an idea of how U.S. students' understanding of geometry concepts compares with that of students from other countries.

The next level outlined by the van Hieles centers on informal deduction. At this level, students formulate definitions and justify relationships among shapes. They are able to explain, for example, why all squares are rectangles but not all rectangles are squares. Venn diagrams are a valuable tool to help students at this level organize their thinking. Those who fail to reach this level of reasoning have difficulty moving to the formal deduction found in high school (van Hiele 1999). In accord with this developmental model, students in grade 3 need explorations that move them from the visual to the descriptive level. As they move from grade 3 to grade 5, they should be given investigations that develop their informal deductive skills. These types of investigations should continue in grades 6–8, prior to exposure to a formal deductive system in high school.

Geometry explorations require tools, and so the classroom should be equipped with a variety of manipulatives such as pattern blocks, geoboards, geoblocks, attribute blocks, tangrams, Legos, protractors, compasses, graph paper, and rulers. Electronic tools should also be available, including geometry software such as Terrapin Logo (2001), Shape Up! (Arita 1995), and TesselMania! Deluxe (1995), and computer games that foster spatial sense such as Tetris™ and The Factory (Cappo and Fish 1992). In addition, the NCTM Illuminations Web site, referred to several times in this book, provides excellent interactive activities—such as analyzing the properties of two- and three-dimensional shapes—to illustrate the big ideas of geometry.

The geometry curriculum for grades 3–5 is based on four big ideas: shapes, location, transformations, and spatial visualization. Students should develop a working geometry vocabulary as they progress through the grades. Precision and clarity in definitions emerge, and the focus should always be on the accuracy of the mathematics involved. Moving students toward the use of sophisticated vocabulary should be a goal. For example, whereas third graders may speak of a *square corner*, fourth graders should describe the same attribute as a *right angle*. In grade 3, students discuss *flips*, but by grade 5, they should speak of *reflections*. The authors have emphasized the focus on vocabulary by including the section "Important Geometric Terms," which sometimes includes definitions and diagrams, in each activity. In fact, the first activity, which focuses on using a variety of geometric terms, can be used as a preassessment for the geometry unit.

Shapes

In prekindergarten–grade 2, students begin to sort figures by their shapes and discover what makes a rectangle a rectangle. In grades 3–5, they move to classifying shapes on the basis of their properties. For example, they might define a parallelogram as a four-sided figure with two sets of parallel sides. Then they move to the informal deductive level and begin to use mathematical arguments to compare and contrast properties among shapes. They also progress to relating two- and three-dimensional figures by investigating such questions as How are a rectangular prism and a rectangle alike? How are they different? Students will continue their explorations in grades 6–8 as they test properties and develop logical arguments.

Students in grades 3–5 should experiment with subdividing, combining, and transforming shapes in both two and three dimensions. They might investigate which shapes they could combine to form rectangles. Teachers can enhance students' explorations by using questions that facilitate connections to prior mathematical knowledge:

- How many different figures can you make?
- Explain why the figures are different from one another.

An activity involving three dimensions might require describing the faces of a square pyramid:

- What shape would you create if you replaced the square with a triangle?
- Now that the base has been changed to a triangle, what is the name of the type of pyramid?

Students also can expand on the idea of two shapes being the same, which introduces the concepts of congruence and similarity. The beginning notions of proportional reasoning emerge as students manipulate similar triangles and explore their properties. In grades 6–8, then, students explore proportional reasoning in greater depth.

Location

In prekindergarten–grade 2, students begin exploring location, direction, and distance using vocabulary such as *over*, *under*, *far*, *between*, *left*, and *right*. In grades 3–5, students use more-sophisticated language to describe position. For example, *parallel*, *perpendicular*, *horizontal*, *vertical*, and *intersecting* are used to locate a line relative to a given line. Students should be engaged in problem-solving activities that require finding the shortest path from one destination to another. Through coordinate geometry, students encounter similarity for the first time. All the activities prepare students for investigations in greater depth in grades 6–8, where students will link geometry to algebra through explorations with slopes and relationships between parallel and perpendicular lines.

Transformations

In grades 3–5, investigations of transformations should move beyond the initial explorations of flips, turns, and slides introduced in the primary grades. More-sophisticated vocabulary is used to describe and predict the results of transformations on shapes. The mathematical terms *reflection*, *rotation*, and *translation* should be introduced. Teachers can explore congruence with their students by considering a series of transformations. Symmetry, including line and rotational symmetry, should be investigated, and a working definition of symmetry should be developed. Again, the focus on precision and clarity of mathematical vocabulary is vital for building understanding.

Spatial Visualization

Spatial visualization is perhaps the most natural extension of students' experiences with geometry and is an underlying theme of all the other big ideas of geometry. An understanding of figure-ground perception, perceptual constancy, position in space, visual memory, and visual discrimination all play a role in students' ability to visualize and manipulate geometric shapes. Students should be encouraged to create, reconstruct, and manipulate shapes and images mentally as well as to combine shapes to create different shapes. They should be allowed ample time to explore three-dimensional shapes physically, to build three-dimensional shapes and then proceed to draw them, and to use nets to build three-dimensional shapes from two-dimensional representations. As they progress, they learn to represent three-dimensional shapes in two dimensions and to examine the differences between two- and three-dimensional representations of shapes and figures.

Students enjoy exploring geometry, especially at these grade levels. Connections abound in geometry: Among the four big ideas, for example, shapes are linked to spatial visualization in the exploration of the third dimension and position is linked to transformations in determining the location of shapes in space. Geometry connects to many other areas of mathematics, such as number sense, algebraic reasoning, and measurement. The ability to look at situations visually, geometrically, and analytically makes students better problem solvers. Teachers must help students see these connections by taking an integrated approach to viewing and doing mathematics.

GRADES 3–5

NAVIGATING *through* GEOMETRY

Appendix

Blackline Masters and Solutions

Geodot Paper for Geoboards

Name ____________________

Quadrilateral Pieces

Name ______________________________

Cut out the quadrilateral pieces below. You will use them with some tasks your teacher assigns.

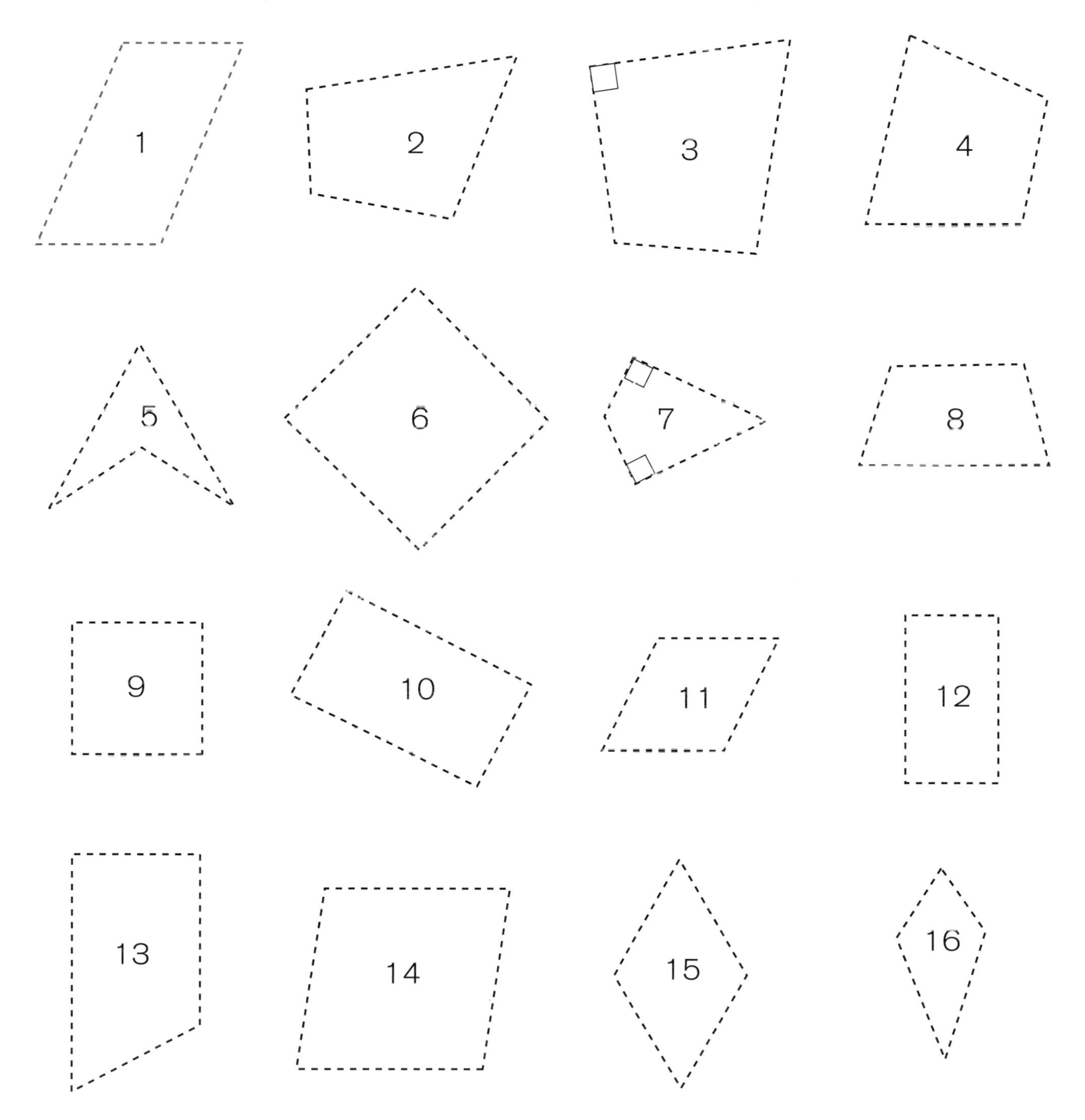

Ring Labels

Name ______________________________

Use hoops or yarn or string to make rings. Then cut out each card for each task, and place it near one of the rings. Place the appropriate quadrilateral pieces in each ring according to the label. You may need to overlap some rings to form intersections.

Task 1

At least one right angle	No right angles

Task 2

All sides the same length	At least one acute angle

Task 3

At least one set of parallel sides	At least one obtuse angle

Task 4

At least two pairs of adjacent sides congruent

All pairs of opposite sides congruent

Task 5 (three rings)

All sides the same length	At least one obtuse angle

At least one right angle

Task 6 (three rings)

At least two pairs of adjacent sides equal

All pairs of opposite angles equal	All adjacent angles equal

Mystery Rings

Name ______________________________

Directions: For each set of mystery rings, make up an appropriate label for each ring and write it above the ring.

Mystery Rings 1

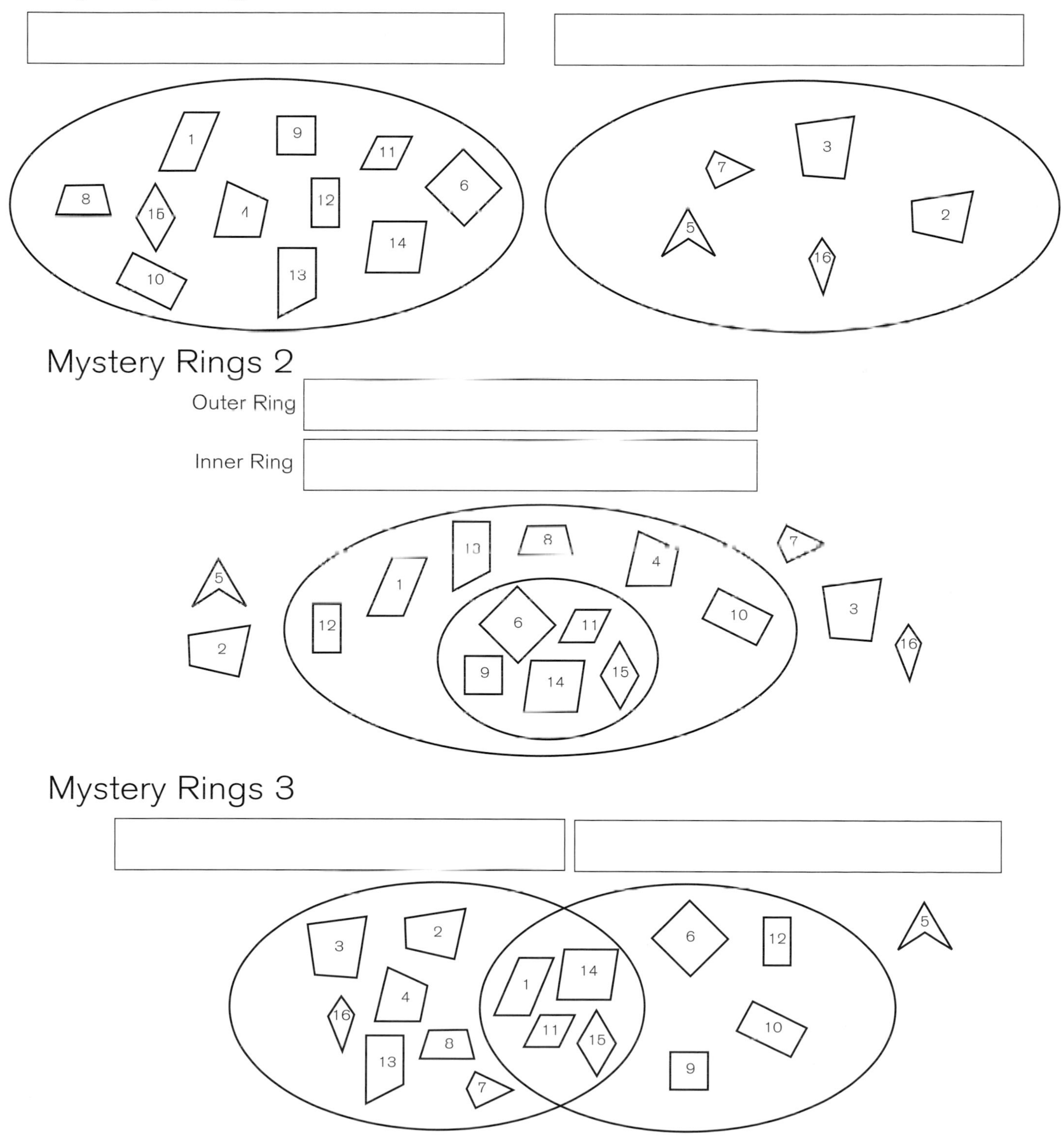

Two- and Three-Dimensional Shapes

Names ______________________________

The three-dimensional shape we chose is a ______________________.

The two-dimensional shape we chose is a ______________________.

Here is a picture of the two-dimensional shape:

List all the ways these two shapes are alike.

List all the ways these two shapes are different.

Counting Parts of Solids

Names ______________________

Figure	Number of Faces	Number of Vertices	Number of Edges

Patterns for the Perfect Solids

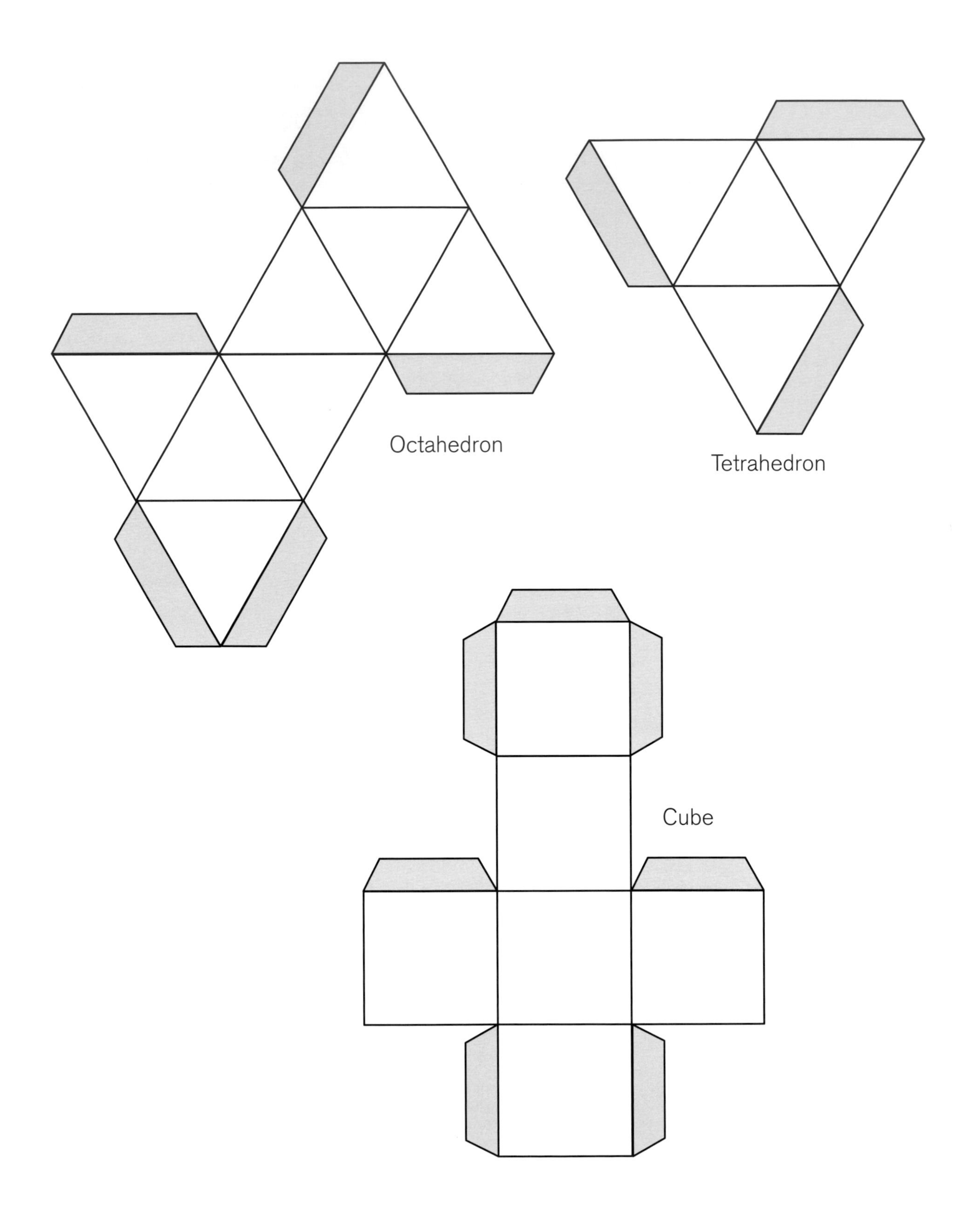

Patterns for the Perfect Solids (continued)

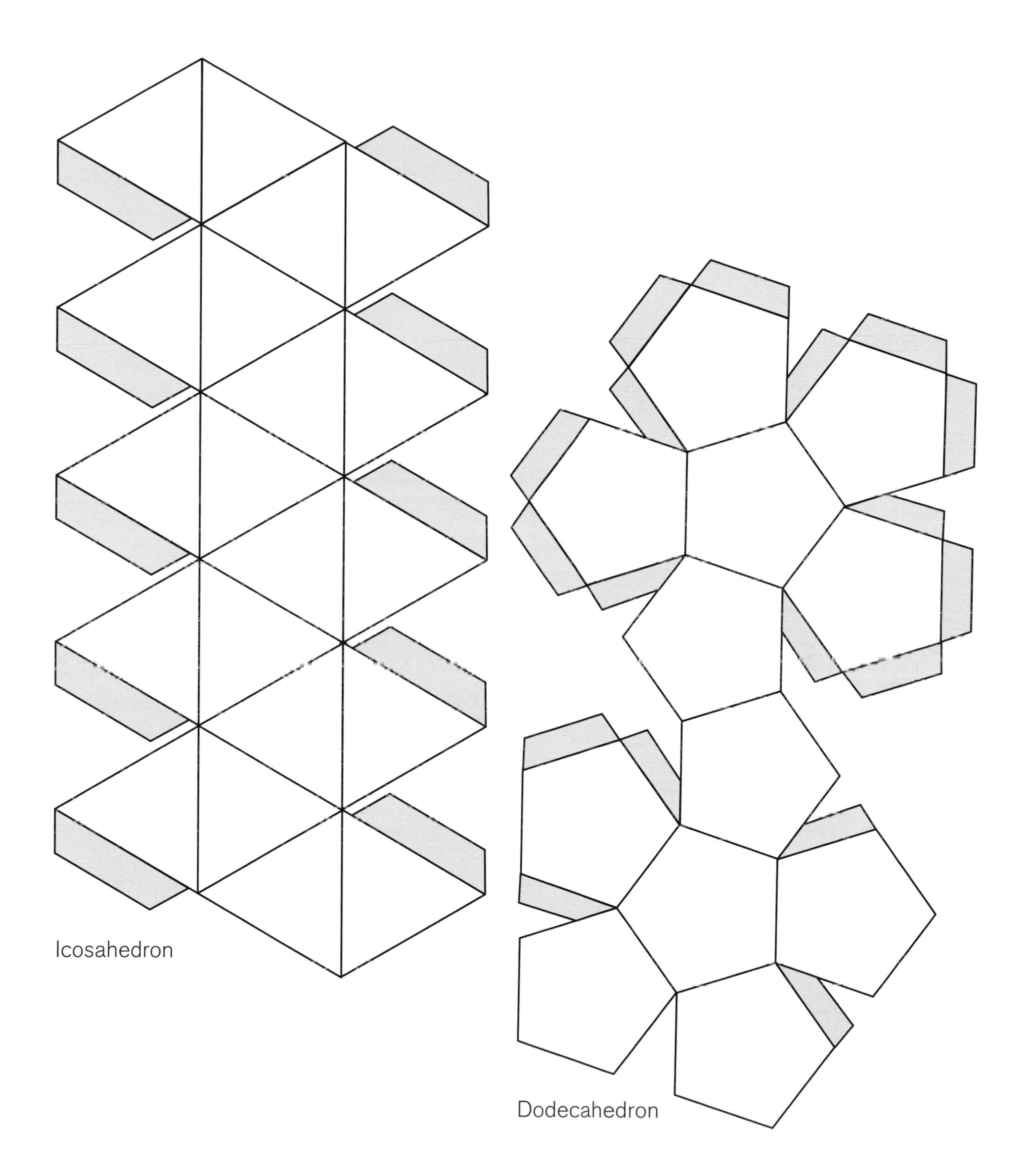

Patterns for Other Solids

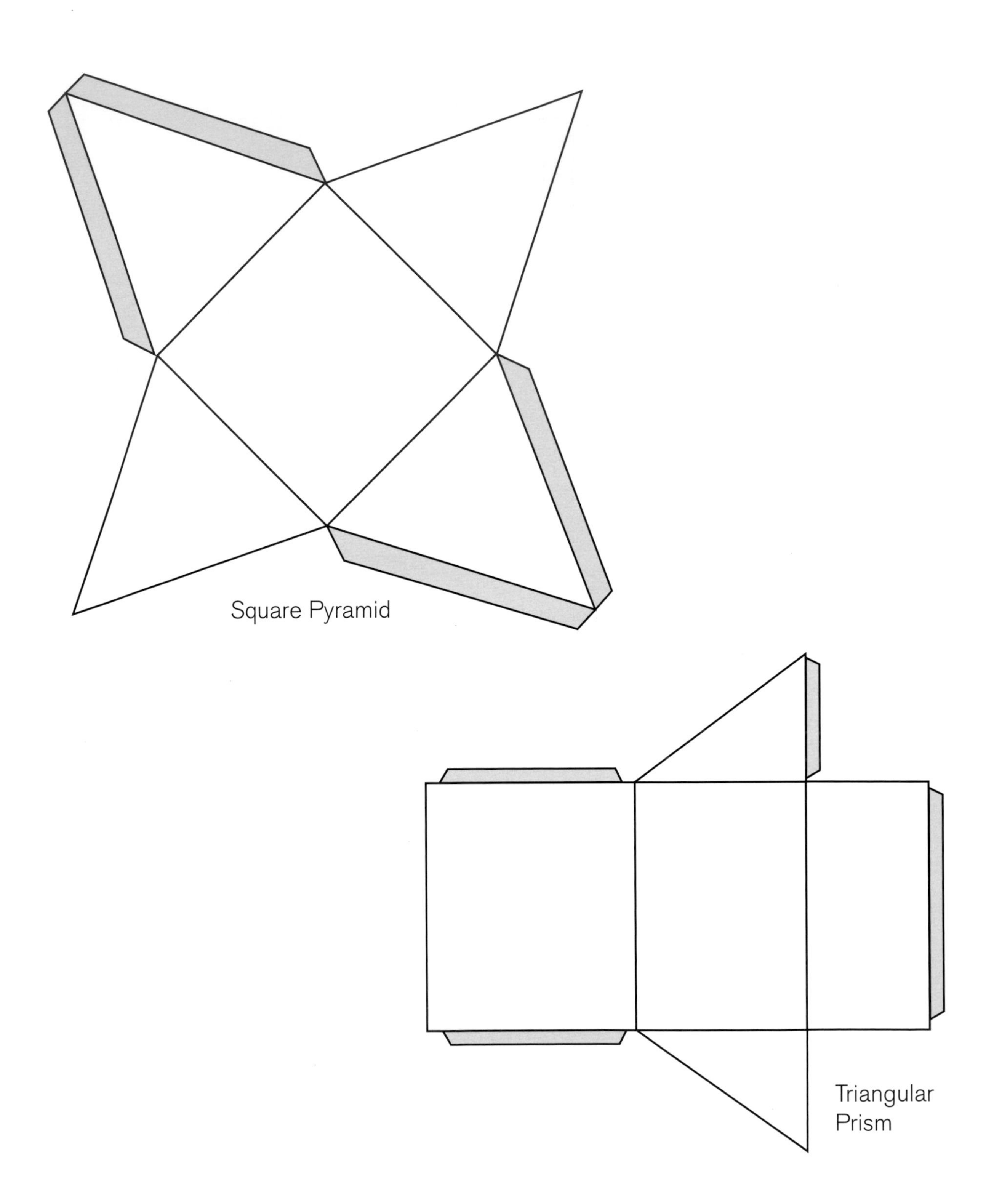

Coordinate Grid A

Name ______________________________

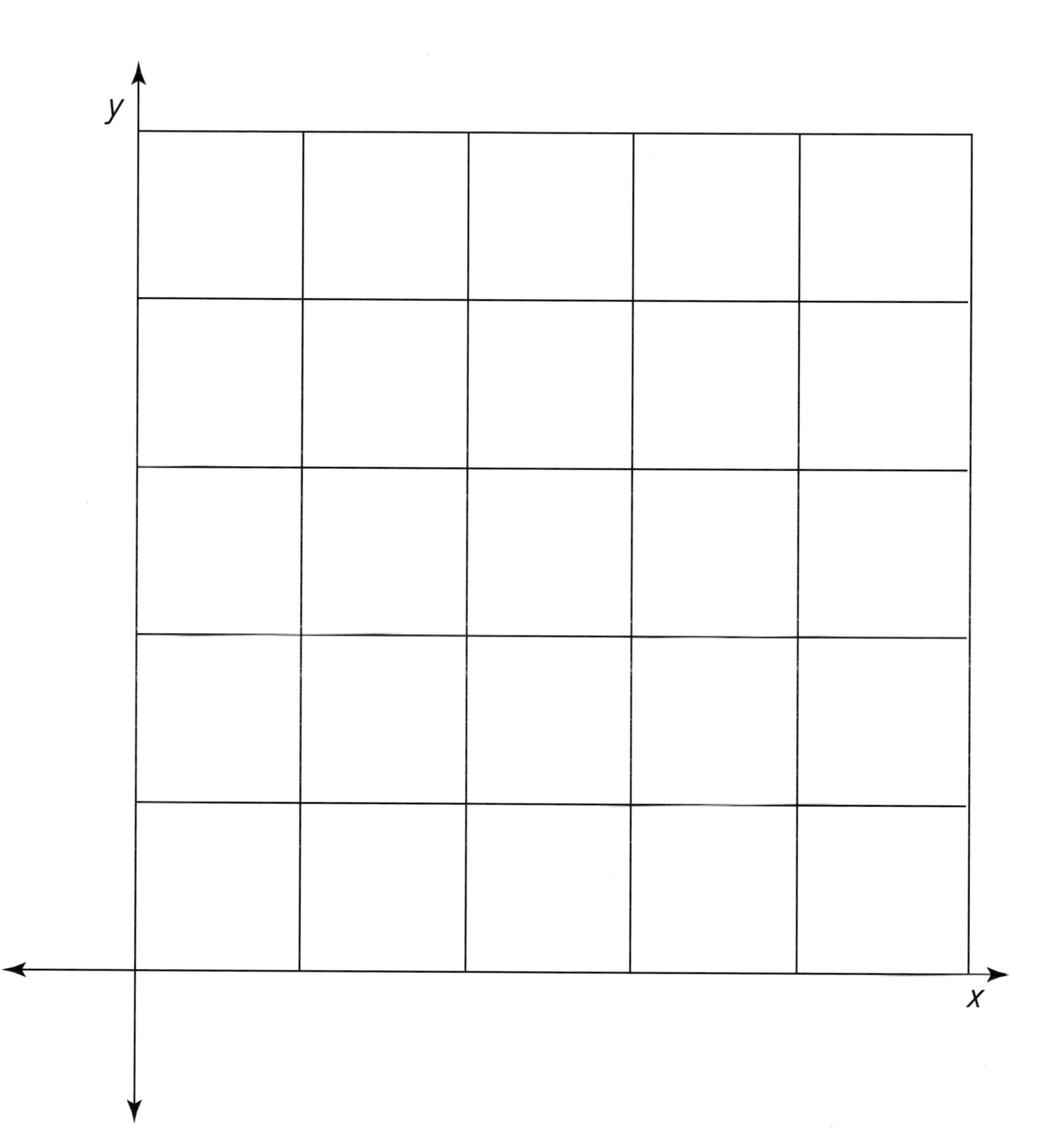

Coordinate Grid B

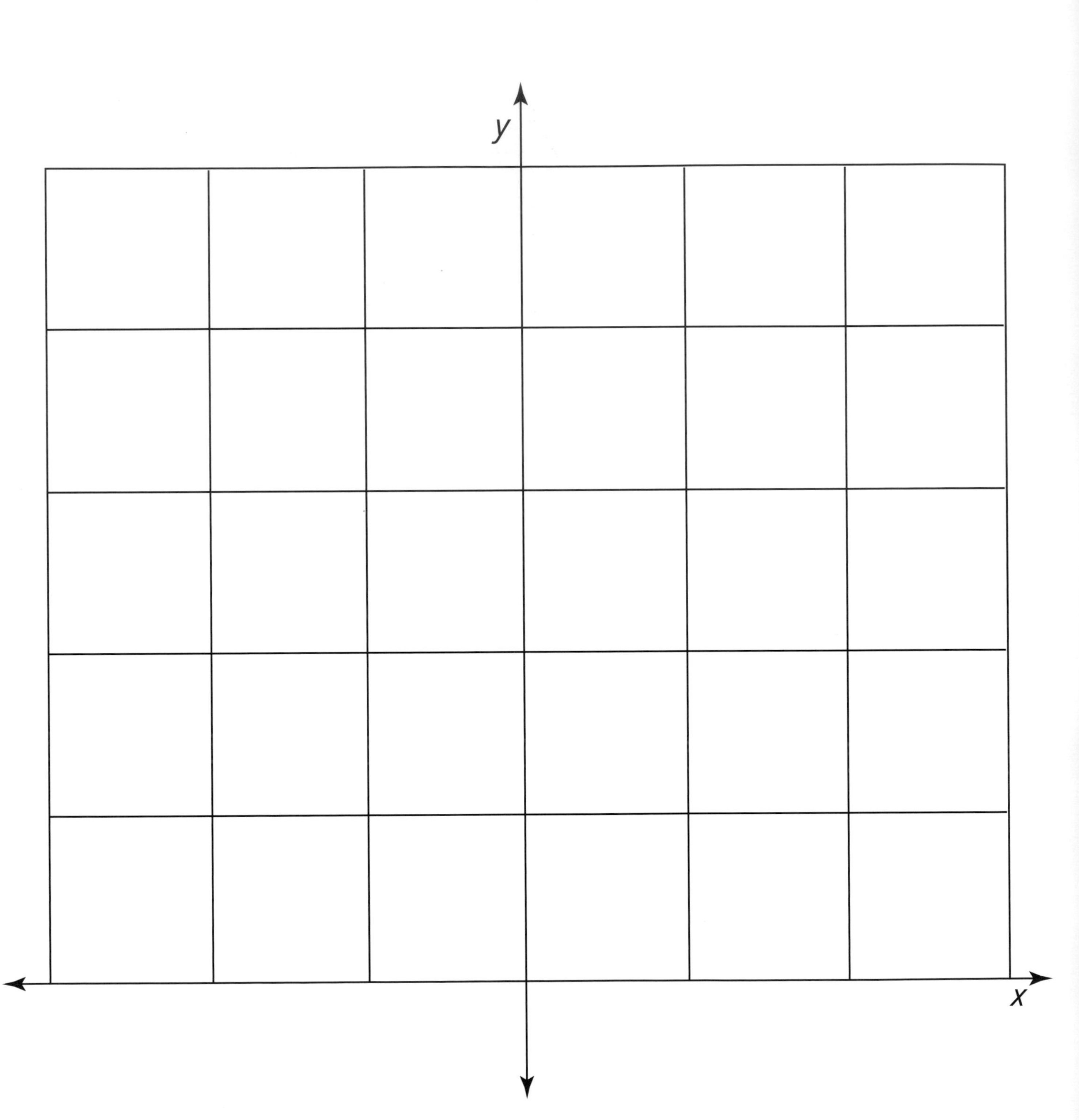

Coordinate Grid C

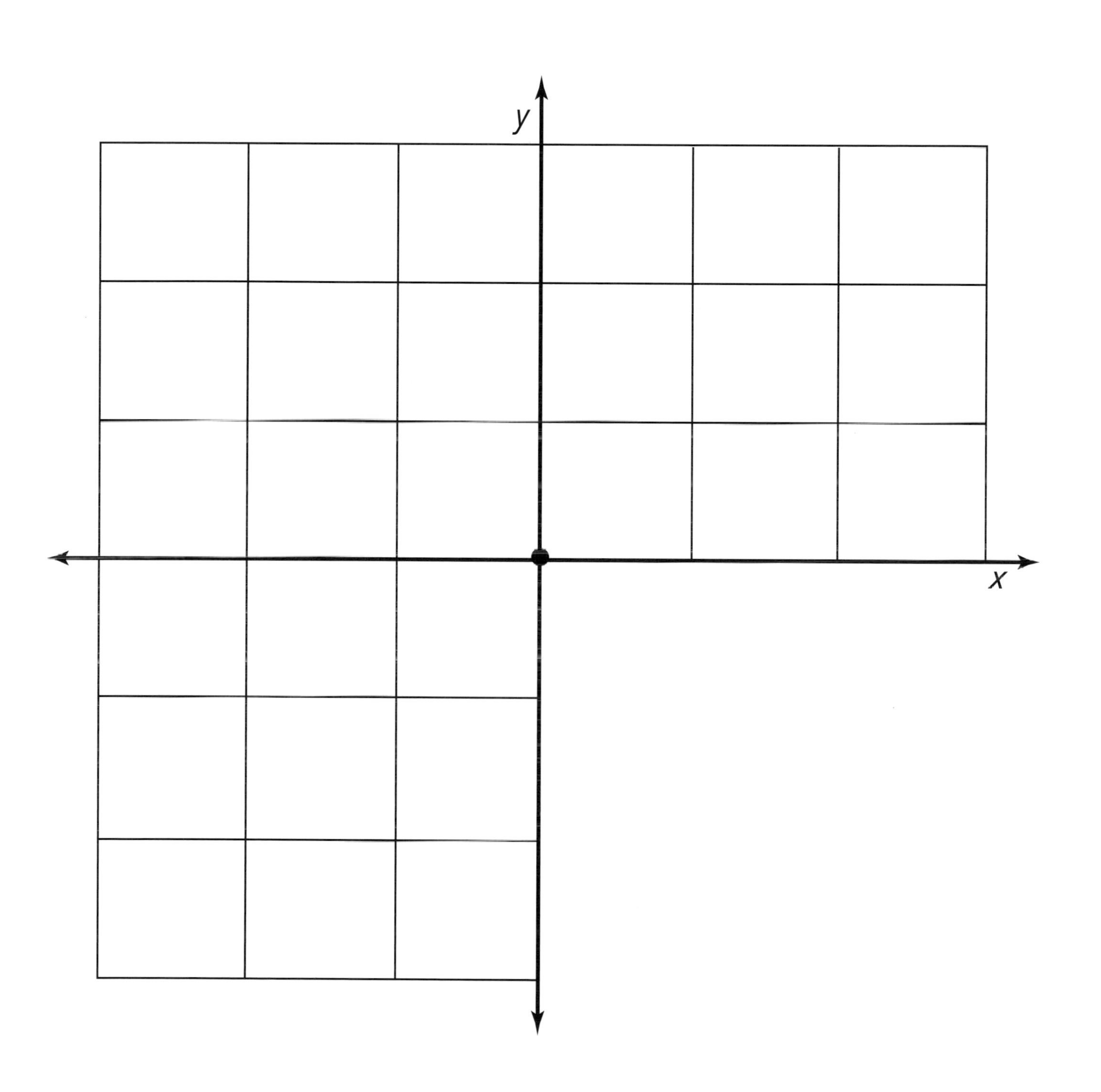

Coordinate Grid D

Name 

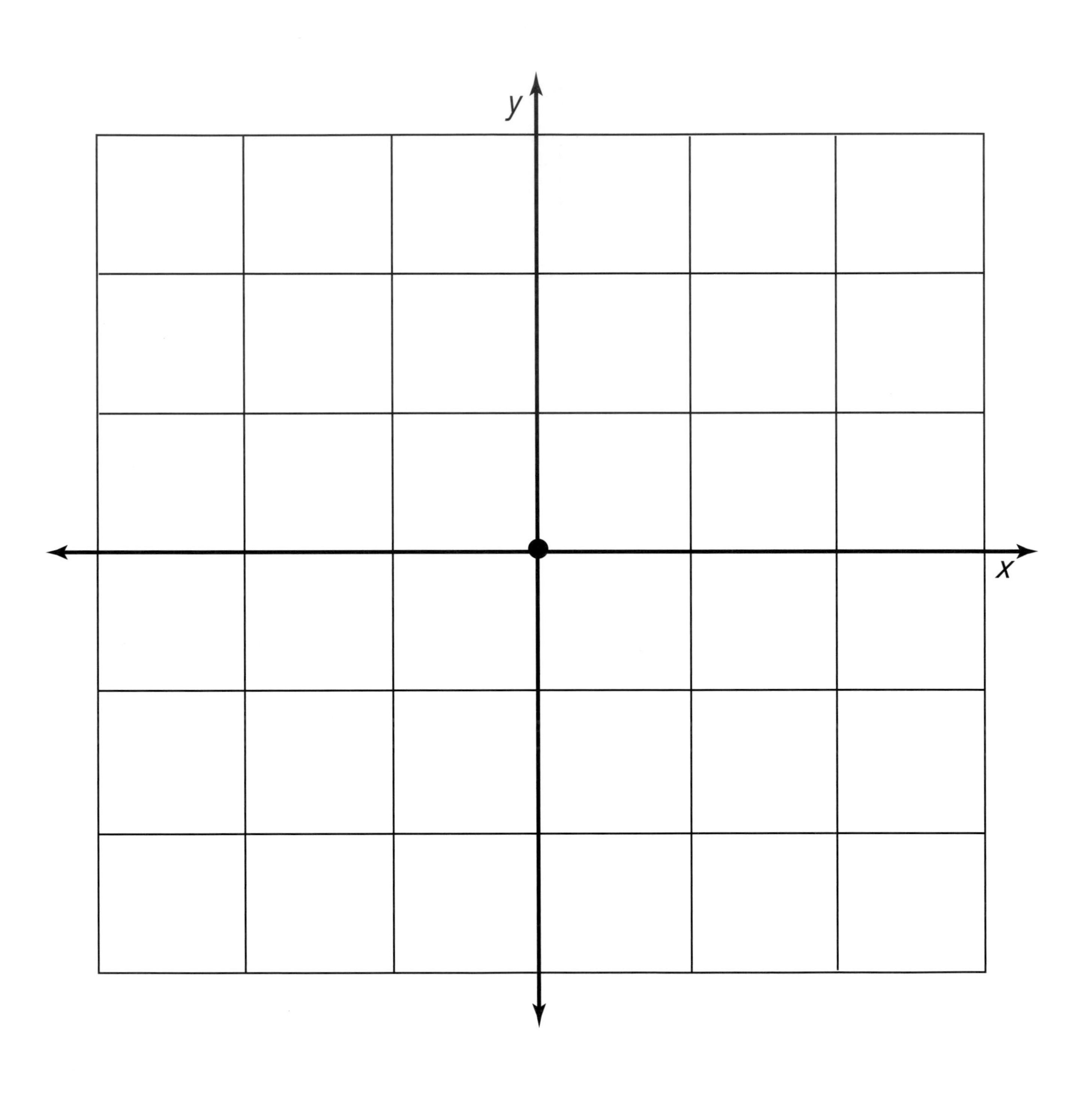

Similar and Nonsimilar Shapes

Name ______________________________

Find the pairs of shapes that are similar. What makes them similar?

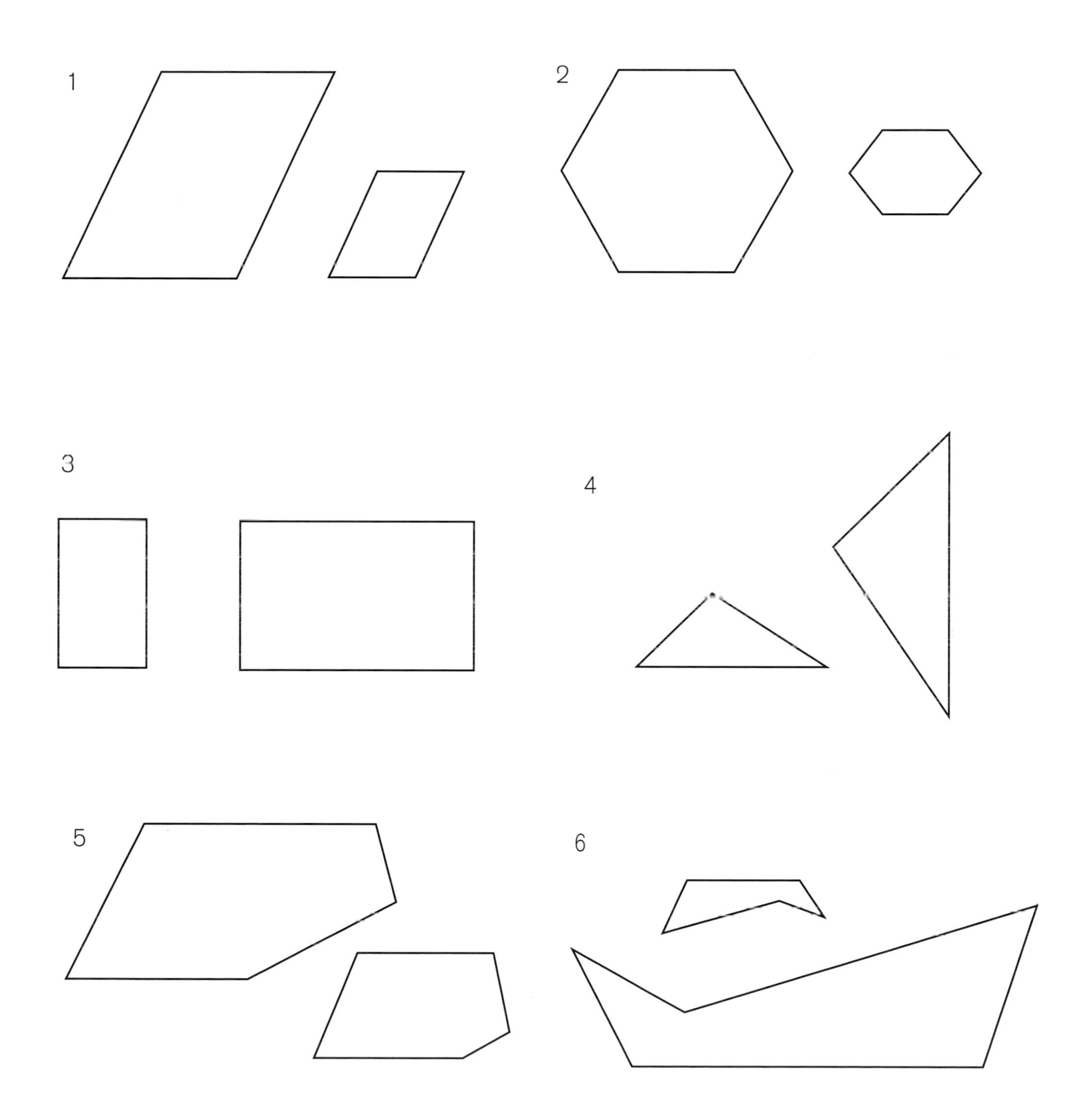

Quarter-Inch Grid Paper

Name ___________________________

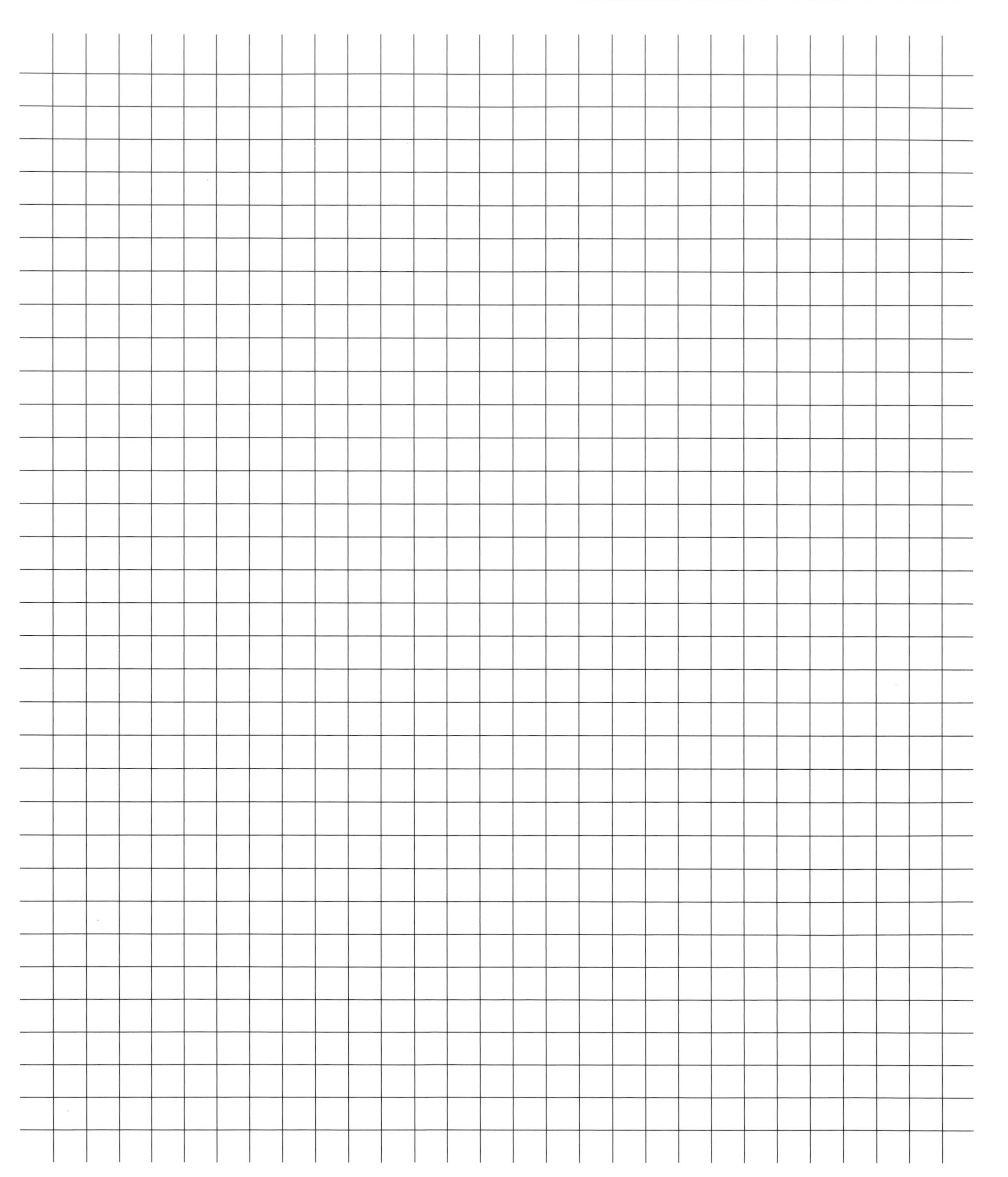

Half-Inch Grid Paper

Name ___________________________________

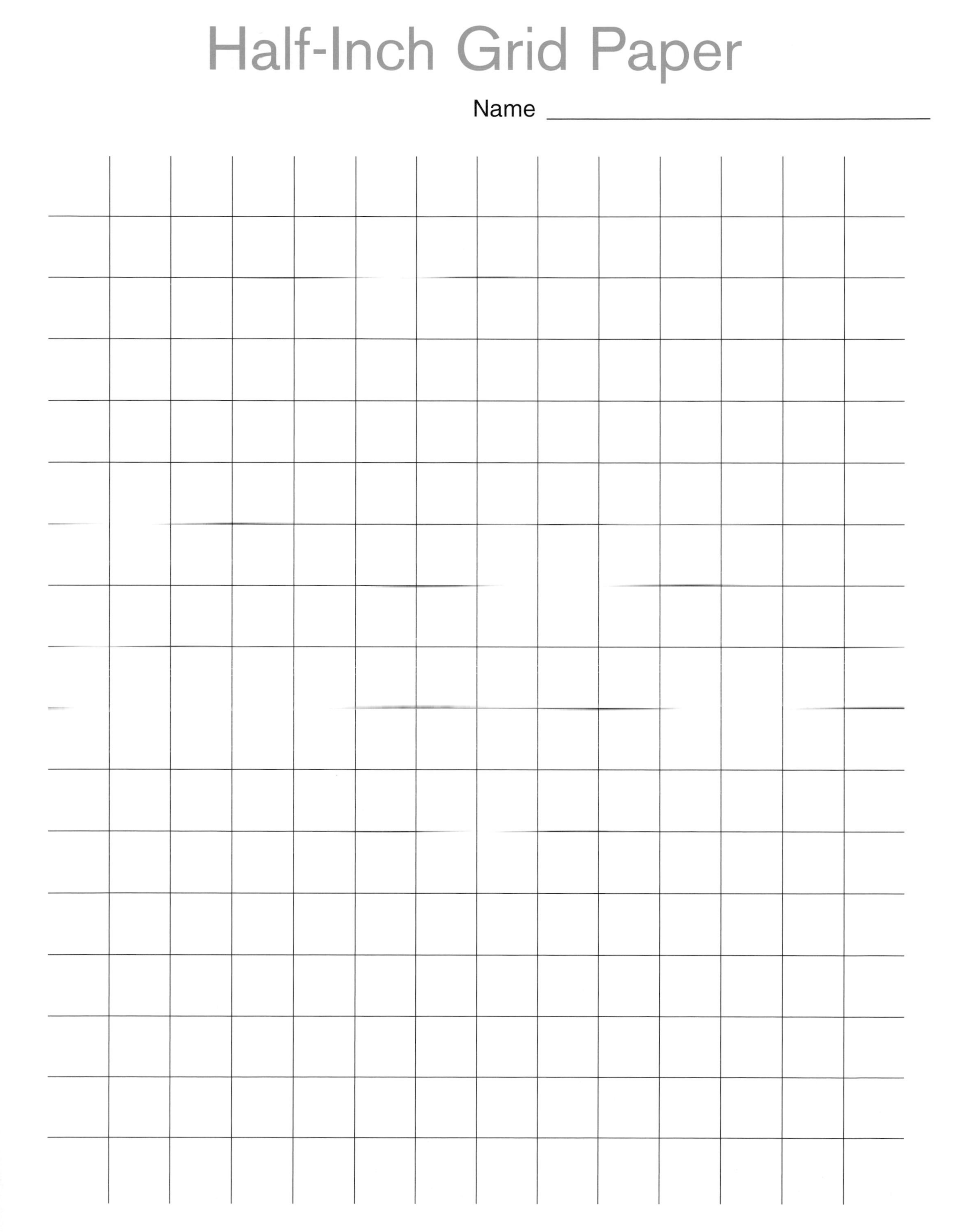

Alphabet Symmetry

Name ______________________________

Draw the lines of symmetry on each of the letters.

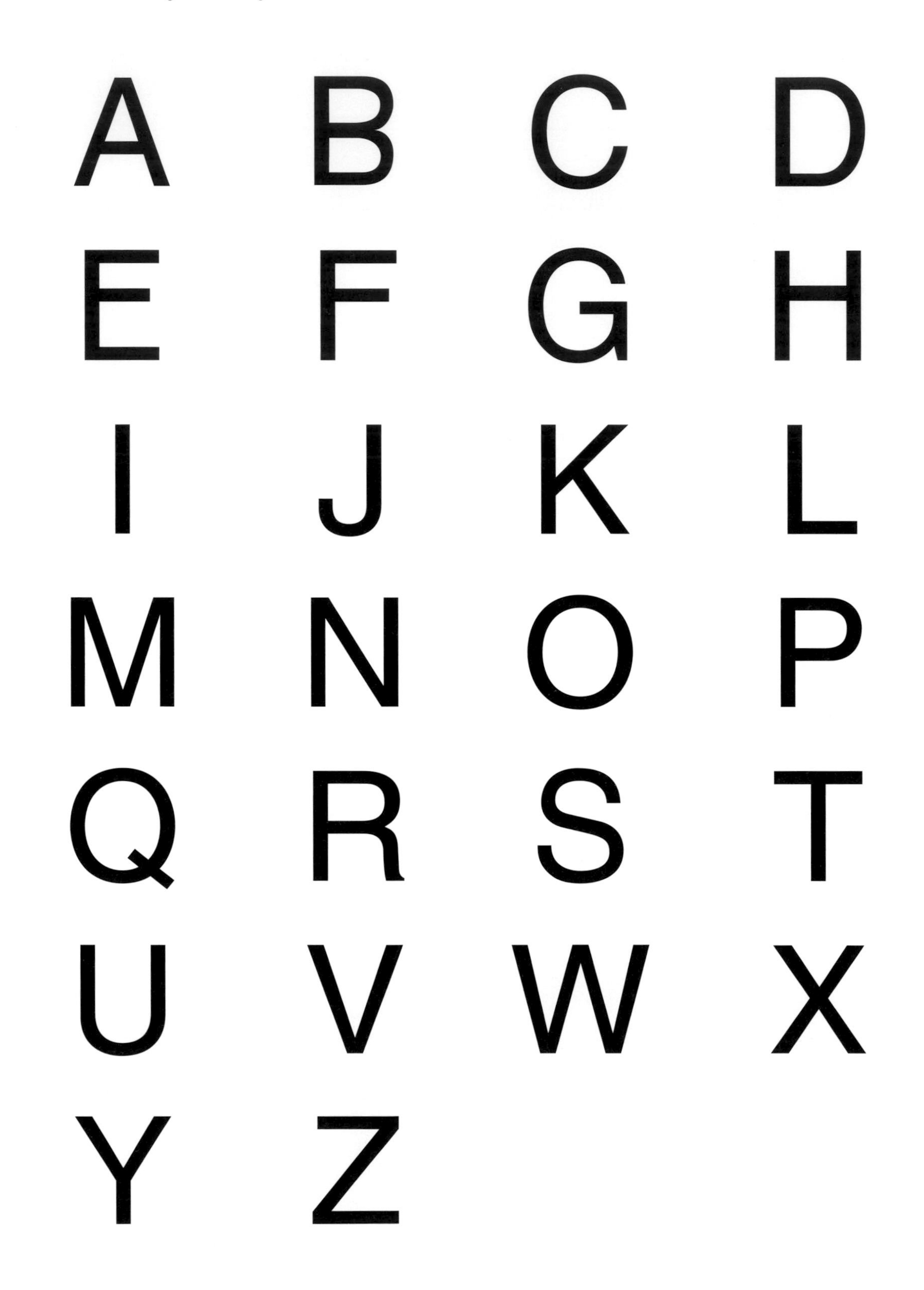

Alphabet-Symmetry Chart

Name ______________________________

Write the letters of the alphabet in the proper column on the chart.

Letters with No Lines of Symmetry	Letters with One Line of Symmetry	Letters with Two Lines of Symmetry	Letters with More Than Two Lines of Symmetry

Paper-Strip Figures

Turn It Around

Names ___________________________

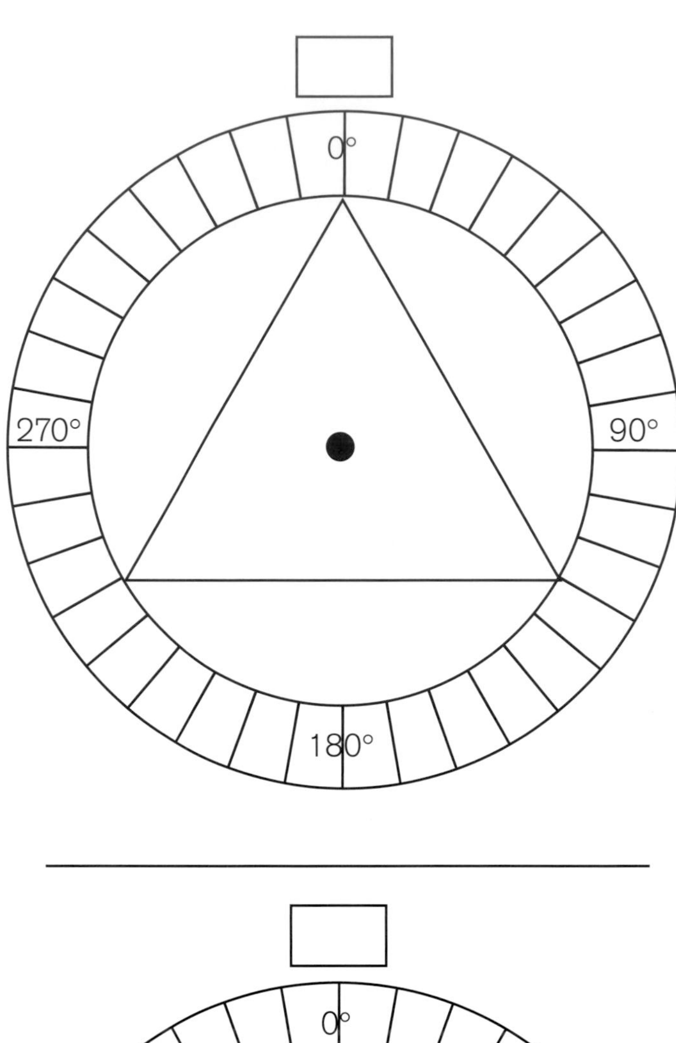

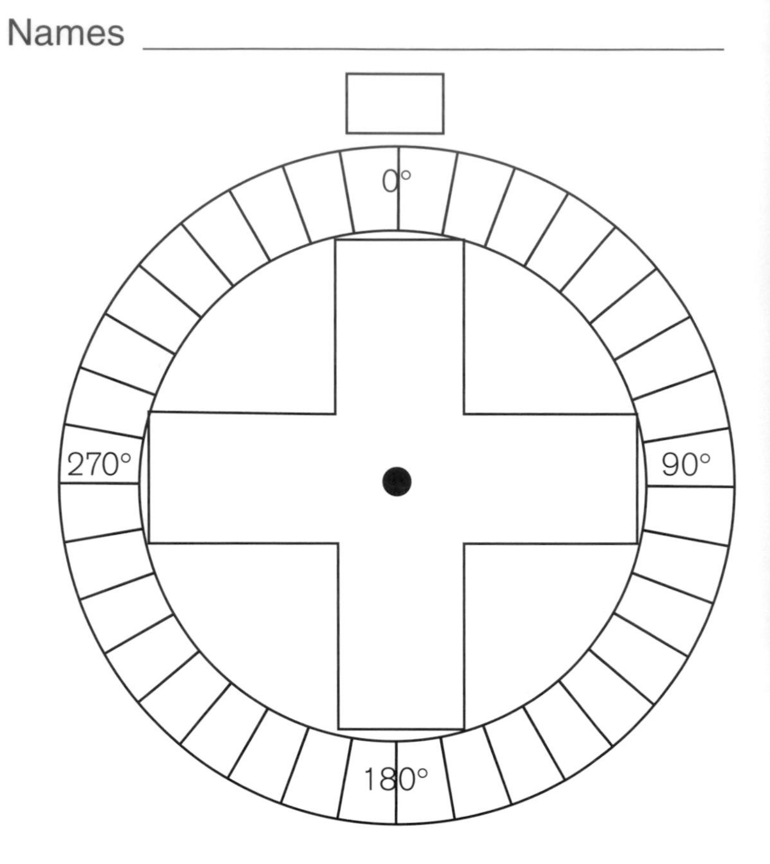

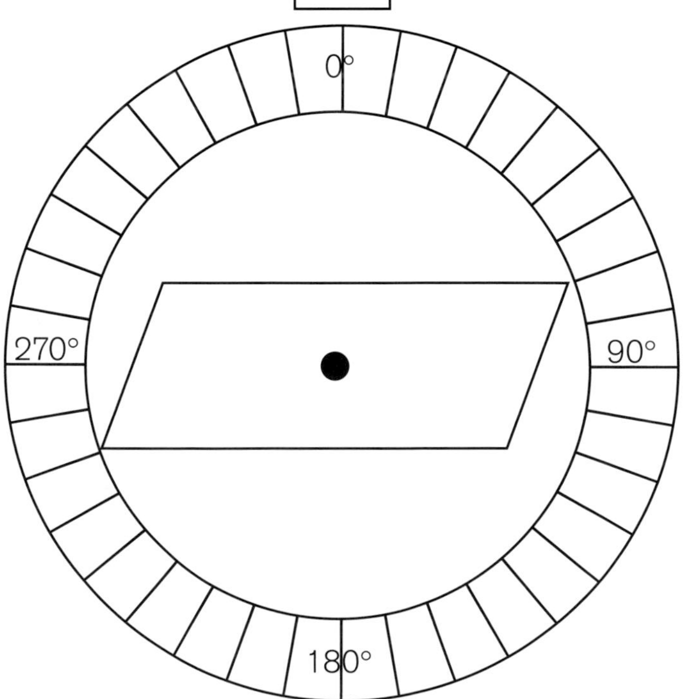

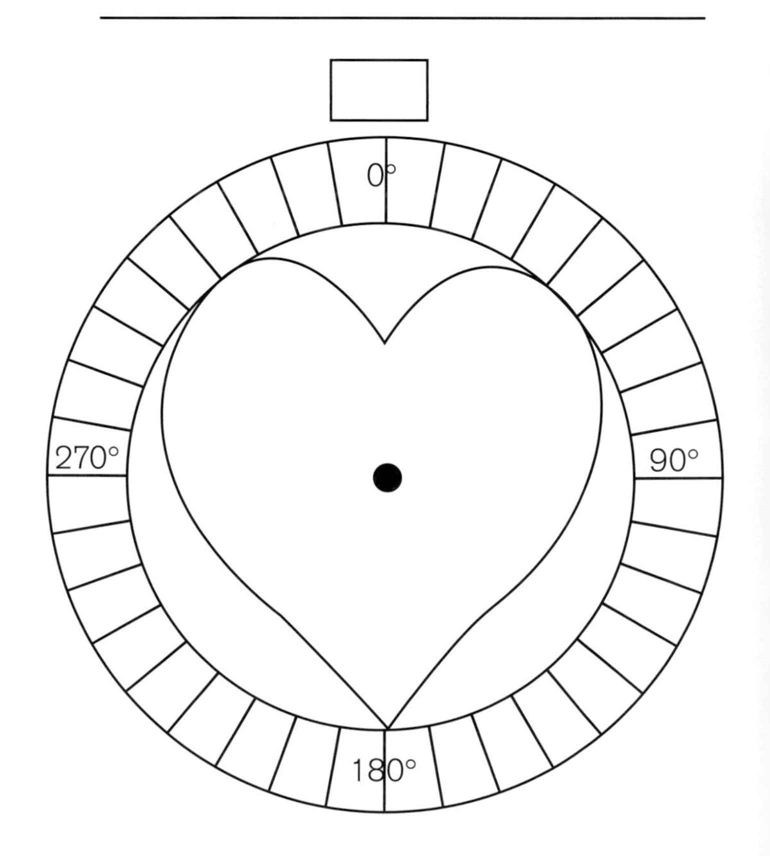

Turn It Around (continued)

Names ______________________

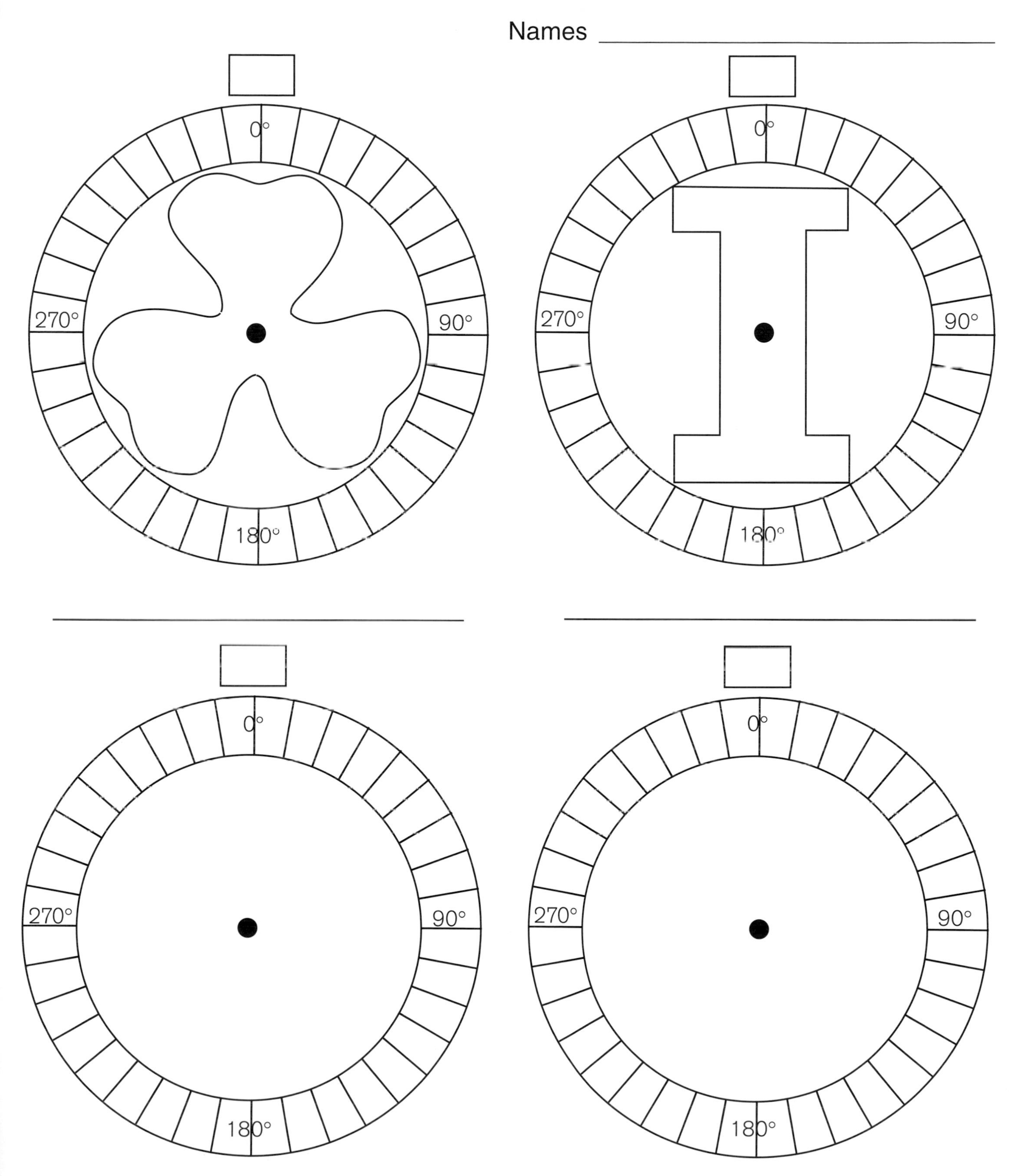

Tetrominoes Cover-Up Game Board

Name ________________________________

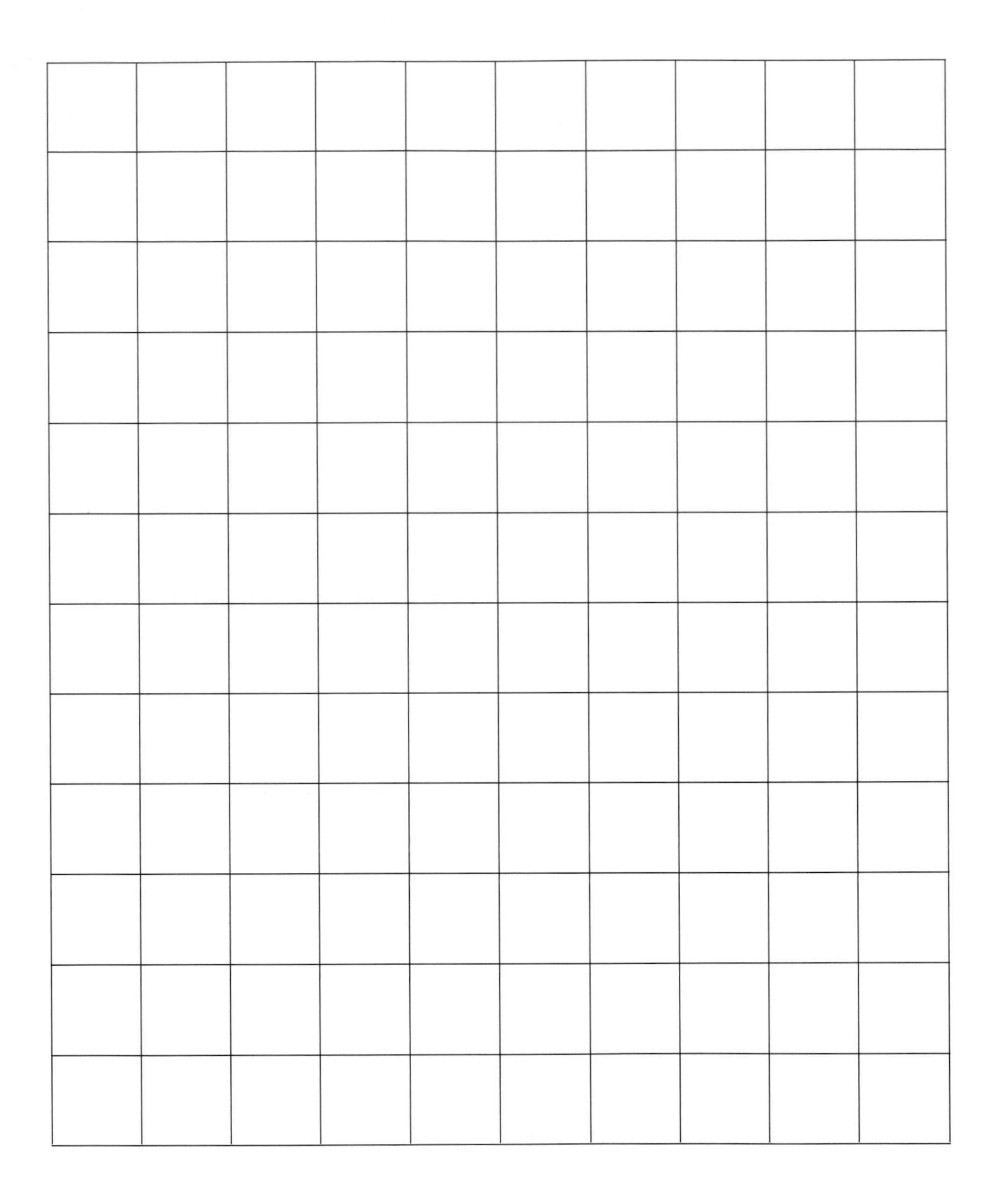

To the Teacher:

This game board is composed of half-inch squares. If the students are playing "tetrominoes cover-up" with pieces made of one-inch squares, copy this grid at 200-percent enlargement onto 11″ × 14″ paper.

Tetrominoes Spinner

Name ____________________

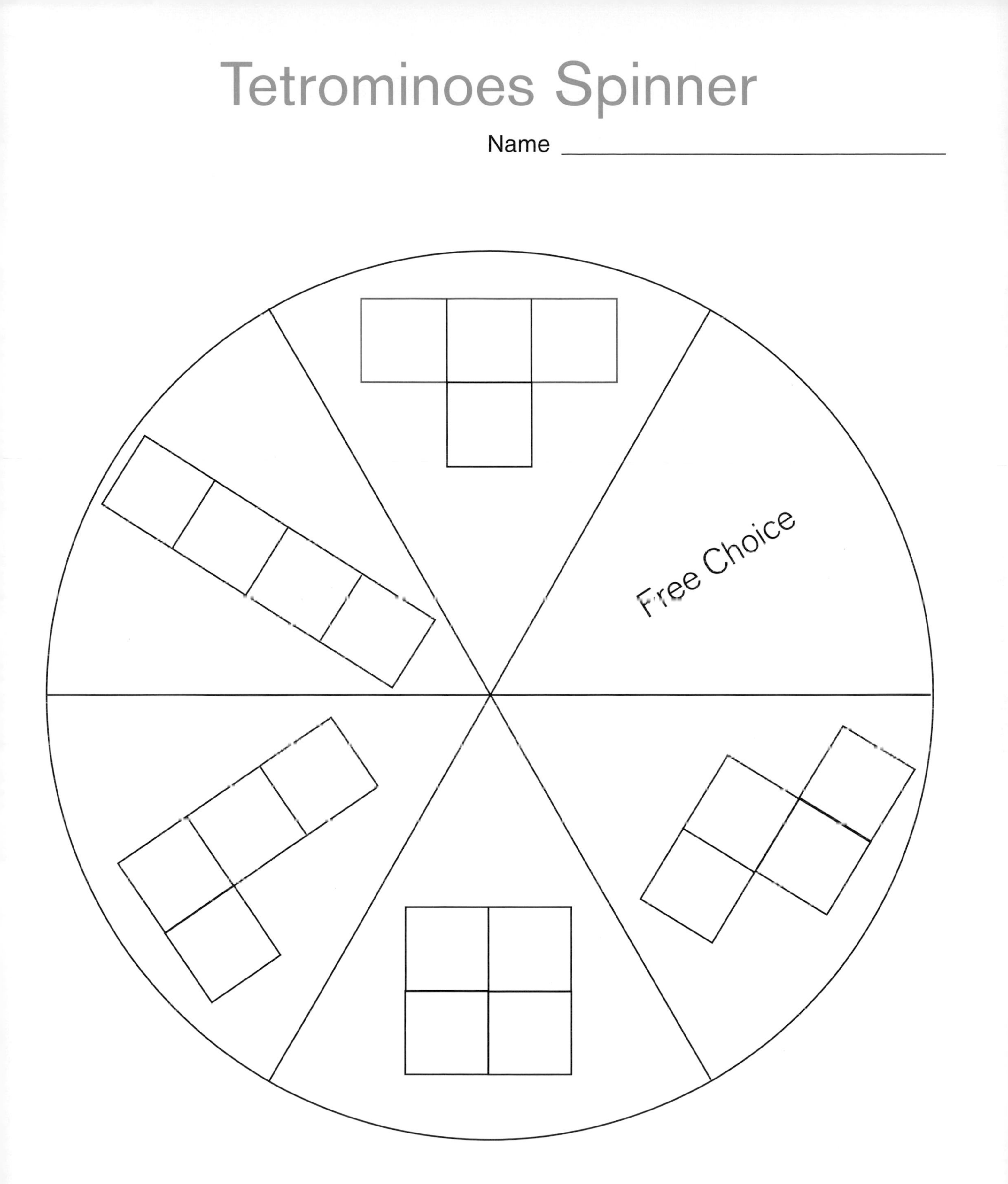

Motion Commotion

Name ______________________________

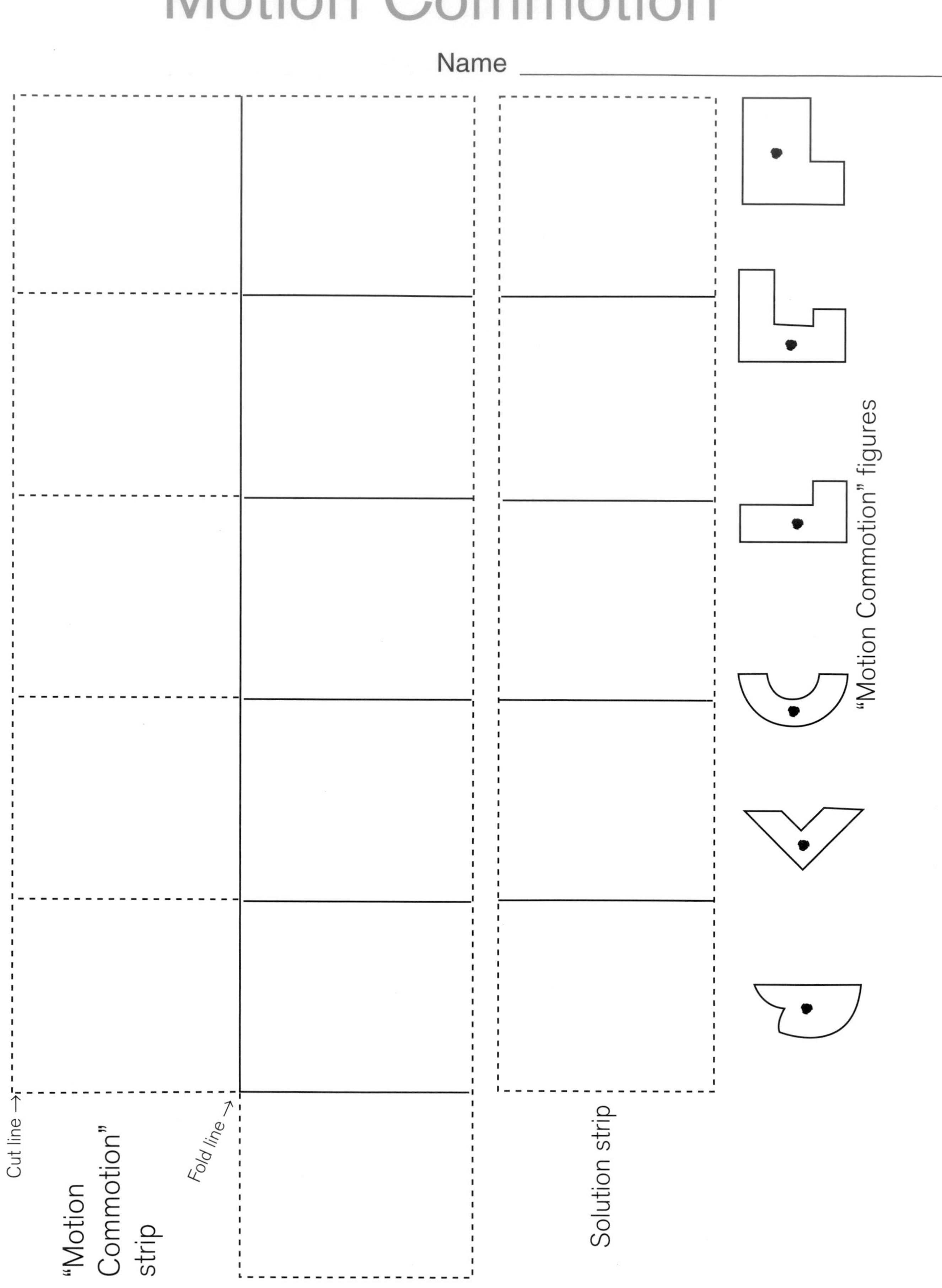

Tangrams

Name ______________________________

Puzzles

Name ______________________________

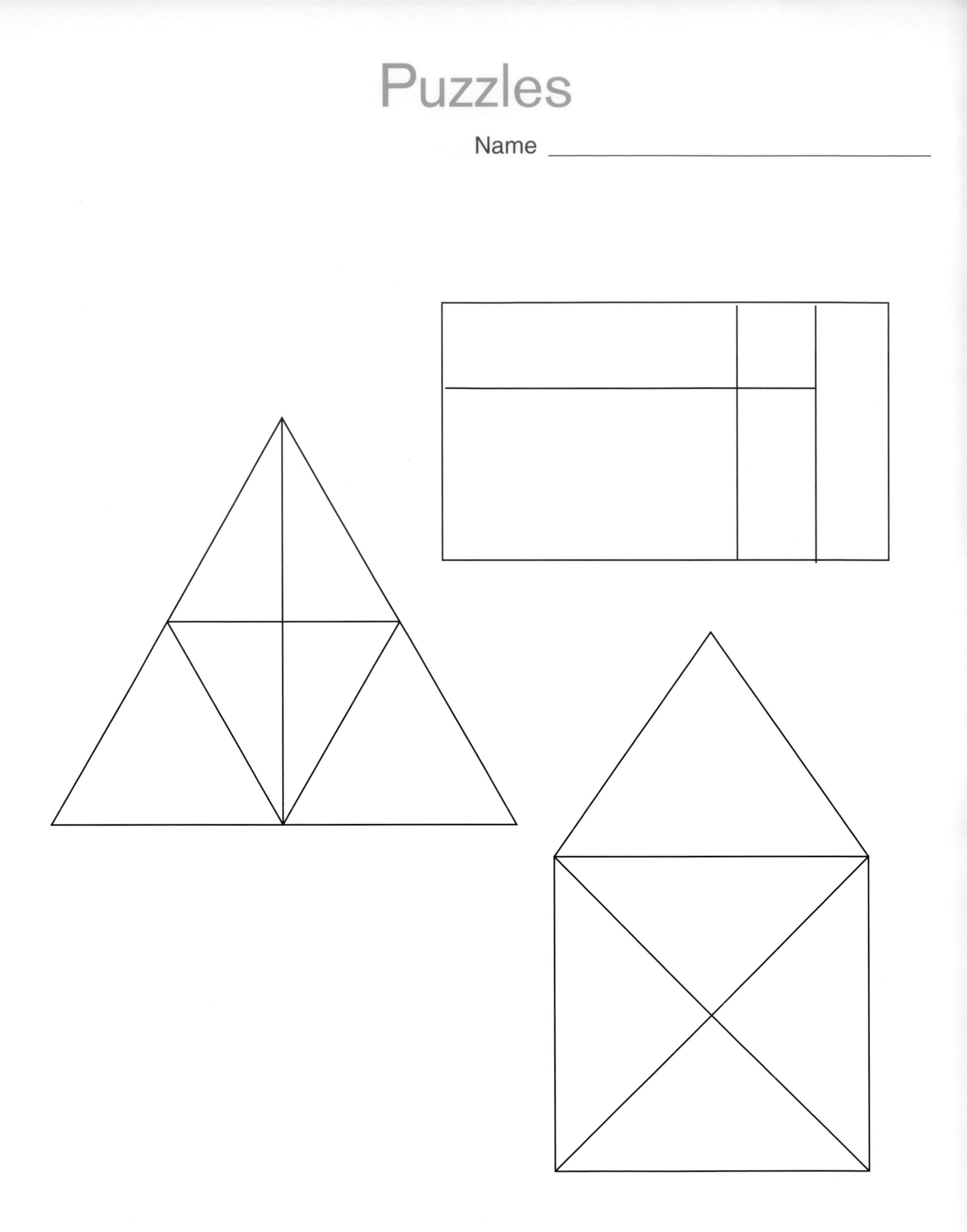

Isodot Paper

Name ______________________________

Geodot Paper

Name ______________________________

One-Inch Grid Paper

Name ______________________________

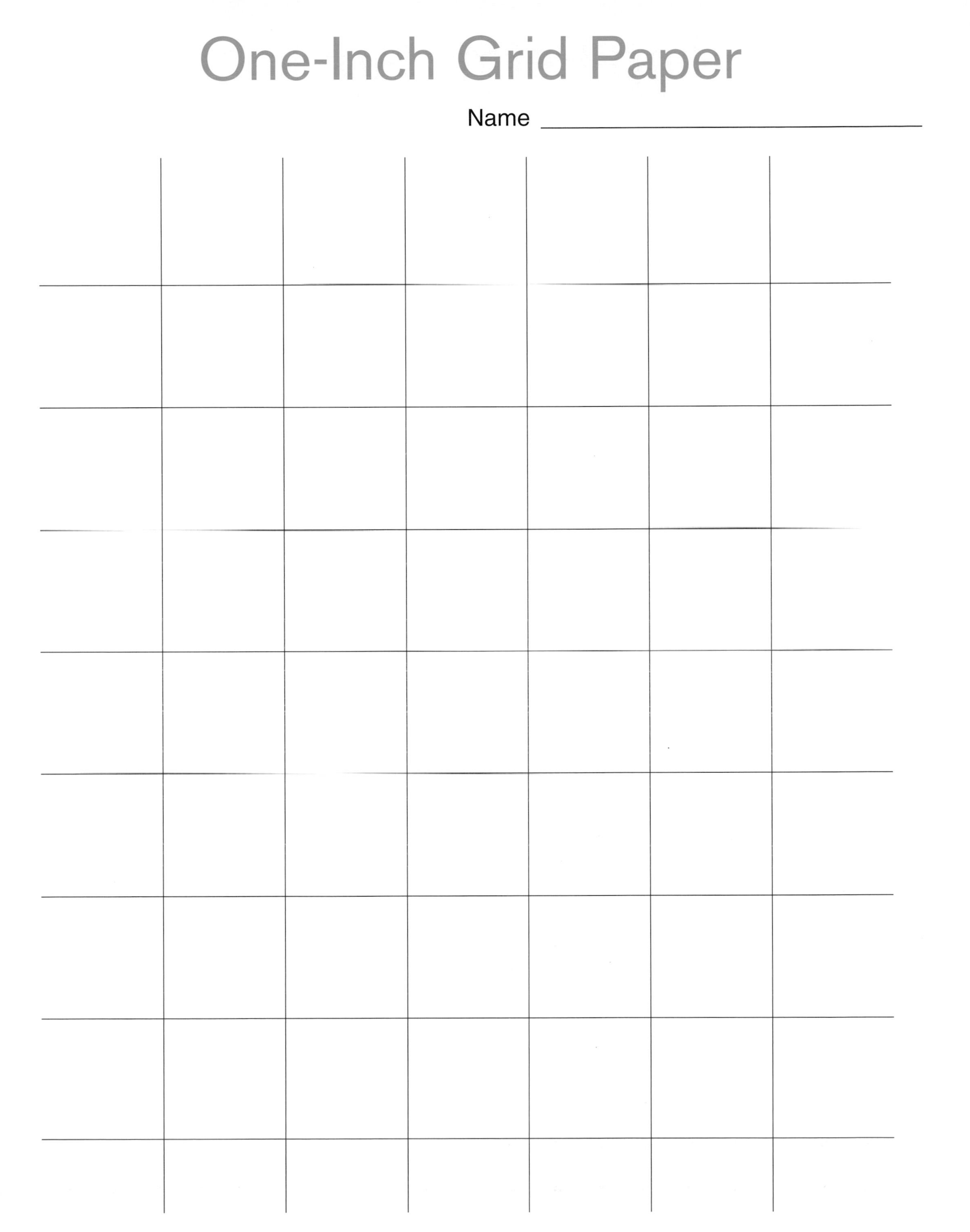

Building Permit Application

Date Submitted ______________________________ Date Approved ____________________

Junior City Planner ____________________________ Building _________________________

Junior City Planner ____________________________ Building _________________________

Junior City Planner ____________________________ Building _________________________

Junior City Planner ____________________________ Building _________________________

City Planner's Approval____

Required Team Building: ____

Please use the space below to indicate the location of the buildings that your team has chosen to build for your city block. Include any parking lots, driveways, sidewalks, streetlights, or other amenities that your block may need. Attach the blueprints of your buildings to this application.

Sample Geo City Map

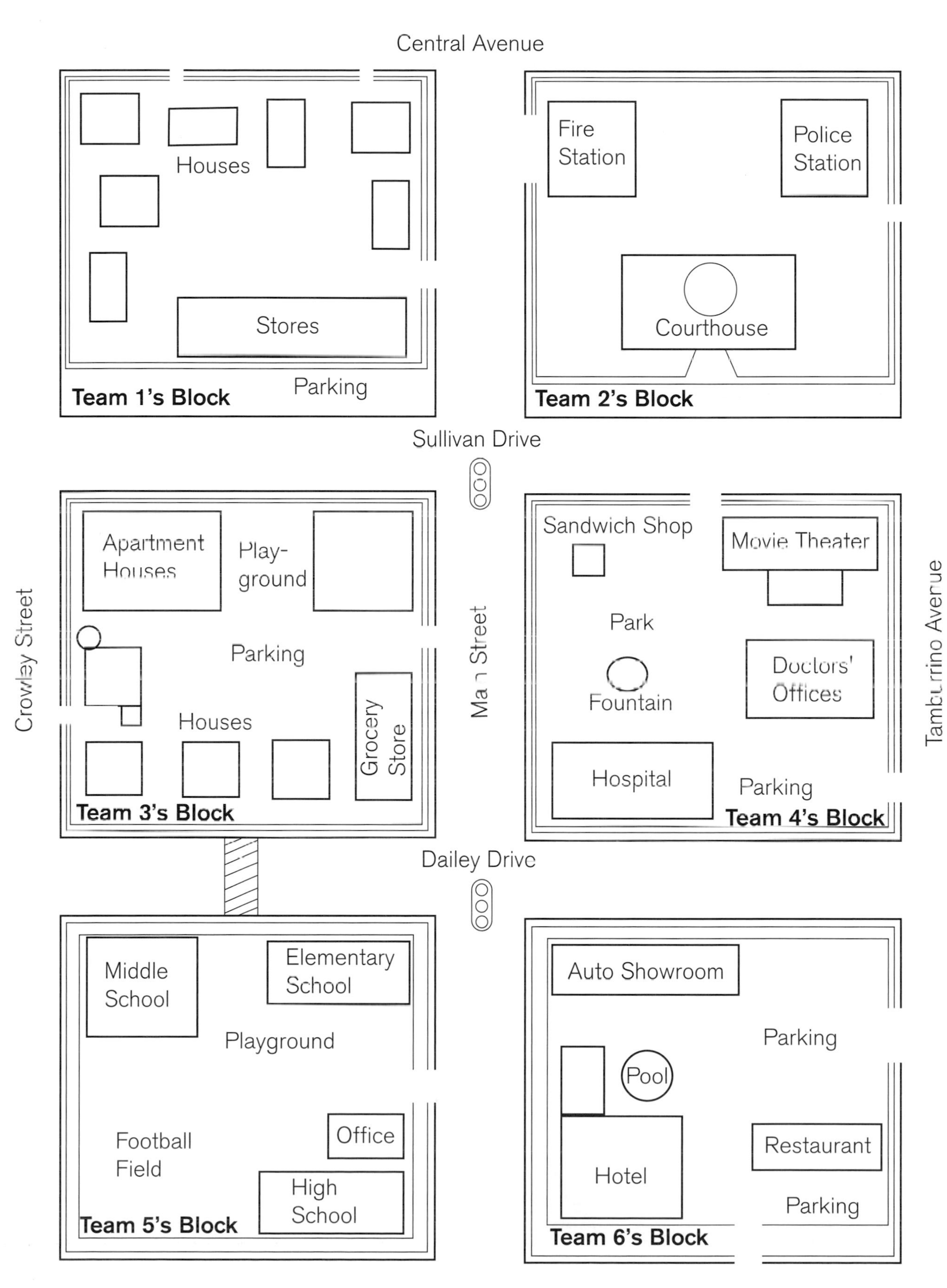

Solutions to Blackline Masters

Possible Solutions to "Mystery Rings"

Mystery Rings 1, left: At least one pair of parallel sides; right: No sides parallel

Mystery Rings 2, inner ring: All sides the same length; outer ring: At least one pair of parallel sides

Mystery Rings 3, left: At least one obtuse angle; right: Both pairs of opposite angles equal

Solutions to "Similar and Nonsimilar Shapes"

Shapes 1, 4, and 6 are similar. Shapes 2, 3, and 5 are nonsimilar.

Solutions to "Alphabet Symmetry"

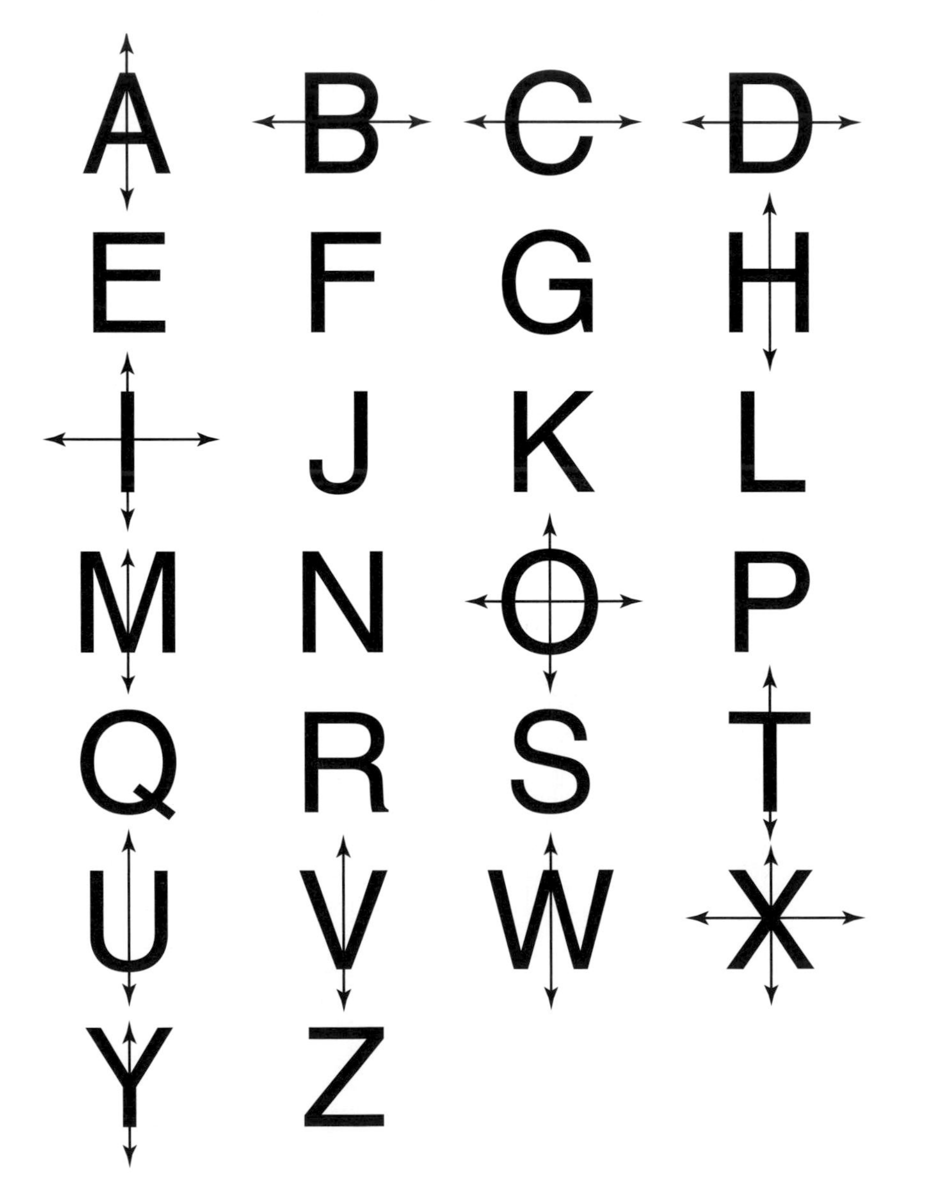

Solutions for "Ring Labels"

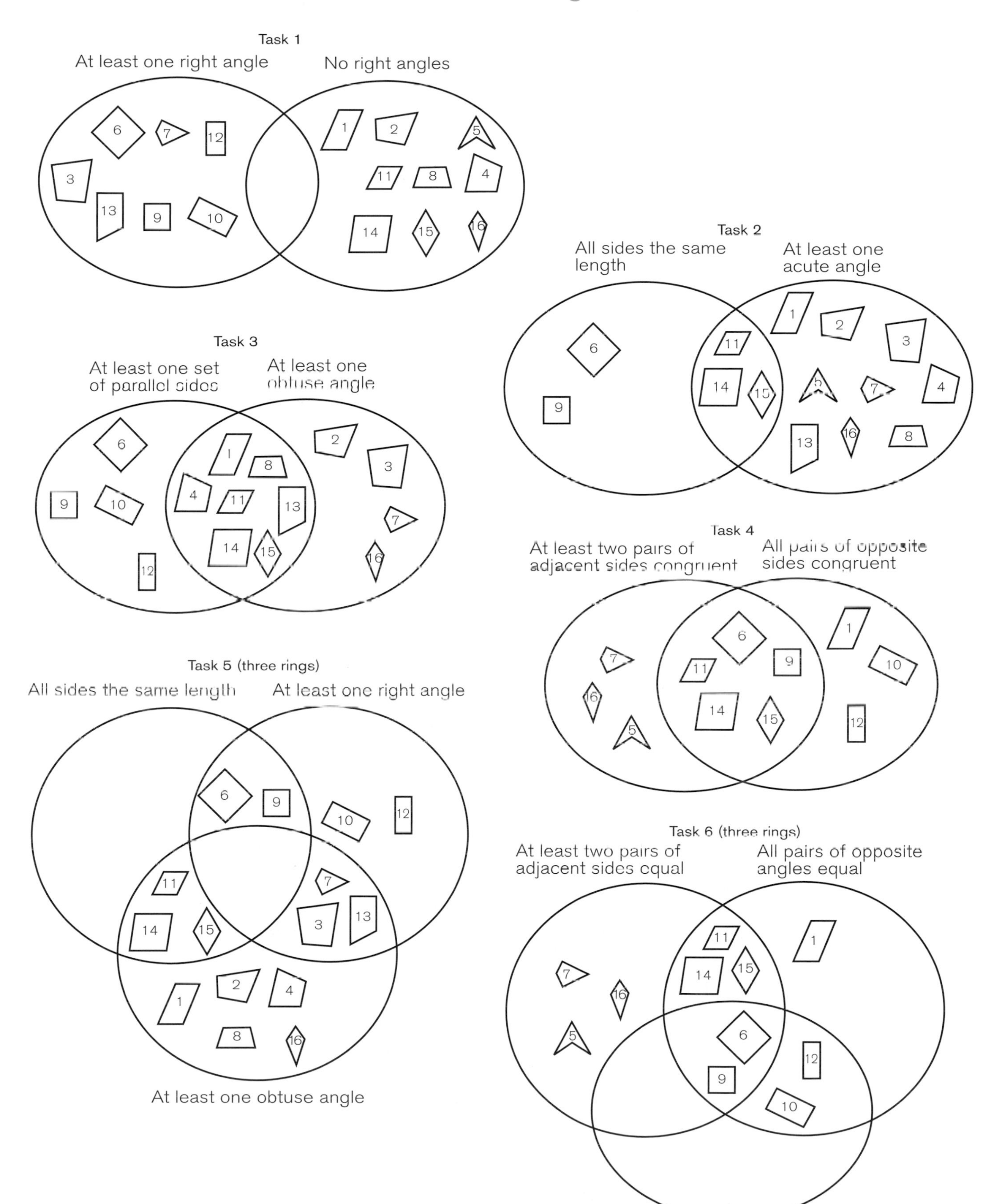

Solutions to "Alphabet-Symmetry Chart"

Letters with No Lines of Symmetry	Letters with One Line of Symmetry	Letters with Two Lines of Symmetry	Letters with More Than Two Lines of Symmetry
E F	A B	I	
G J	C D	O	
K L	H M	X	
N P	T U		
Q R	V W		
S Z	Y		

Solutions to "Paper-Strip Figures"

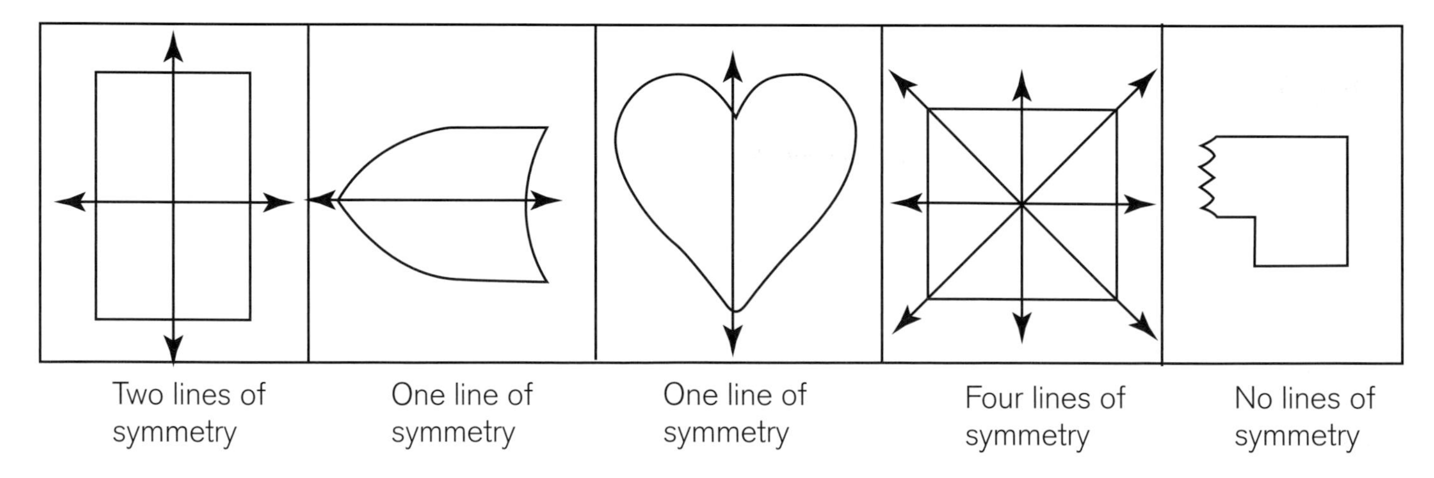

Two lines of symmetry | One line of symmetry | One line of symmetry | Four lines of symmetry | No lines of symmetry

References

Arita, Eleanor. Shape Up! Pleasantville, N.Y.: Sunburst Technology, 1995.

Battista, Michael T. "Geometry Results from the Third International Mathematics and Science Study." *Teaching Children Mathematics* 5 (February 1999): 367–73.

Battista, Michael T., and Douglas H. Clements. "Exploring Solids and Boxes." In *Investigations in Number, Data, and Space.* Menlo Park, Calif.: Dale Seymour Publications, 1998.

Burger, William F., and J. Michael Shaughnessy. "Characterizing the van Hiele Levels of Development in Geometry." *Journal for Research in Mathematics Education* 17 (January 1986): 31–48.

Cappo, Marge, amd Mike Fish. The Factrory. Pleasantville, N.Y.: Sunburst Communications, 1992.

Clements, Douglas H., Susan Jo Russell, Cornelia Tierney, Michael T. Battista, and Julie Sarama. "Flips, Turns, and Area: 2-D Geometry." In *Investigations in Number, Data, and Space.* Menlo Park, Calif.: Dale Seymour Publications, 1998.

Dahlstrom, Carol Field, ed. *One Hundred One Full-Size Quilt Blocks and Borders.* Des Moines, Iowa: Better Homes and Gardens, 1998.

Downie, Diane, Twila Slesnick, and Jean Kerr Stenmark. *Math for Girls and Other Problem Solvers.* Berkeley, Calif.: Lawrence Hall of Science, University of California, 1981.

Ernst, Lisa Campbell. *Sam Johnson and the Blue Ribbon Quilt.* New York: Lothrop, Lee & Shepard Books, 1983.

"Find the Picture." In *Ideas from the "Arithmetic Teacher," Grades 4–6*, compiled by Francis (Skip) Fennell and David Williams, pp. 73–74. Reston, Va.: National Council of Teachers of Mathematics, 1986.

Flournoy, Valerie. *The Patchwork Quilt.* New York: Dial Books for Young Readers, 1985.

Giganti, Paul Jr., and Mary Jo Cittadino. "The Art of Tessellation." *Arithmetic Teacher* 37 (March 1990): 6–16.

Granger, Tim. "Math Is Art." *Teaching Children Mathematics* 7 (September 2000): 10–13.

Hopkins, Mary Ellen. *The It's Okay If You Sit on My Quilt Book.* Santa Monica, Calif.: ME Publications, 1989.

Hopkinson, Deborah. *Sweet Clara and the Freedom Quilt.* New York: Dragonfly Books, 1993.

Huse, Vanessa Evans, Nancy Larson Bluemel, and Rhonda Harris Taylor. "From Paper to Pop-Up Books." *Teaching Children Mathematics* 1 (September 1994): 14–17.

Irvine, Joan. *How to Make Pop-Ups.* Mansfield Center, Conn.: Creative Learning Press, 1987.

———. *How to Make Super Pop-Up Books.* Mansfield Center, Conn.: Creative Learning Press, 1992.

Jackiw, Nick. The Geometer's Sketchpad. Version 3.0. Emeryville, Calif.: Key Curriculum Press, 1991.

Lehrer, Richard, and Carmen L. Curtis. "Why Are Some Solids Perfect? Conjectures and Experiments by Third Graders." *Teaching Children Mathematics* 6 (January 2000): 324–29.

National Council of Teachers of Mathematics (NCTM). *Principles and Standards for School Mathematics.* Reston, Va.: NCTM, 2000.

Oberdorf, Christine D., and Jennifer Taylor-Cox. "Shape Up!" *Teaching Children Mathematics* 5 (February1999): 340–45.

Reimer, Wilbert, and Luetta Reimer. *Historical Connections in Mathematics.* Vol. 1. Fresno, Calif.: AIMS Education Foundation, 1992.

Renshaw, Barbara S. "Symmetry the Trademark Way." *Arithmetic Teacher* 34 (September 1986): 6–12.

Schifter, Deborah. "Learning Geometry: Some Insights Drawn from Teacher Writing." *Teaching Children Mathematics* 5 (February 1999): 360–66.

Stenmark, Jean Kerr, Virginia Thompson, and Ruth Cossey. *Family Math.* Berkeley, Calif.: Lawrence Hall of Science, University of California, 1986.

TesselMania! Deluxe. Minneapolis, Minn.: MECC, 1995. Software.

van Hiele, Pierre M. "Developing Geometric Thinking through Activities That Begin with Play." *Teaching Children Mathematics* 5 (February 1999): 310–16.

Walker, Kathryn, Cynthia Reak, and Kelly Stewart. *Twenty Thinking Questions for Geoboards.* Mountain View, Calif.: Creative Publications, 1995a.

———. *Twenty Thinking Questions for Pattern Blocks.* Mountain View, Calif.: Creative Publications, 1995b.

Additional Resources

Battista, Michael T. "The Importance of Spatial Structuring in Geometric Reasoning." *Teaching Children Mathematics* 6 (November 1999):170–77.

Bertheau, Myrna. "The Most Important Thing Is …." *Teaching Children Mathematics* 1 (October 1994): 112–15.

Carroll,William M. "Cross Sections of Clay Solids." *Arithmetic Teacher* 35 (March 1988): 6–11.

Clark, Clare, and Betsy Carter. *Math in Stride.* Menlo Park, Calif.: Addison-Wesley Publishing Co., 1989.

Claus, Alison. "Exploring Geometry." *Arithmetic Teacher* 40 (September 1992): 14–17.

Escher, M. C., and J. L. Locher. *The World of M. C. Escher.* New York: Harry N. Abrams, 1974.

Evered, Lisa J. "Folded Fashions: Symmetry in Clothing Design." *Arithmetic Teacher* 40 (December 1992): 204–6.

The Graphic Work of M. C. Escher. New York: Ballantine Books, 1971.

Juraschek, William. "Get in Touch with Shape." *Arithmetic Teacher* 37 (April 1990): 14–16.

Liedtke, Werner W. "Developing Spatial Abilities in the Early Grades." *Teaching Children Mathematics* 2 (September 1995): 12–18.

Nitabach, Elizabeth, and Richard Lehrer. "Developing Spatial Sense through Area Measurement." *Teaching Children Mathematics* 2 (April 1996): 473–76.

Pereira-Mendoza, Lionel. "Geometry and Language—a Natural Connection." *Teaching Children Mathematics* 3 (April 1997): 454–57.

Prentice, Gerard. "Flexible Straws." *Arithmetic Teacher* 37 (November 1989): 4–5.

Renga, Sherry. "Ideas: The Mail Route." *Arithmetic Teacher* 41 (April 1994): 465–66, 471.

Sakshaug, Lynae. "The Mathematics of Motion." *Teaching Children Mathematics* 6 (December 1999): 250–51.

Schloemer, Cathy G. "Tips for Teaching Cartesian Graphing: Linking Concepts and Procedures." *Teaching Children Mathematics* 1 (September 1994): 20–23.

Seidel, Judith Day. "Symmetry in Season." *Teaching Children Mathematics* 4 (January 1998): 244–49.

Sellke, Donald H. "Geometric Flips via the Arts." *Teaching Children Mathematics* 5 (February 1999): 379–83.

Seymour, Dale, and Jill Britton. *Introduction to Tessellations*. Palo Alto, Calif.: Dale Seymour Publications, 1989.

Shaw, Jean M., Conn Thomas, Ann Hoffman, and Janice Bulgren. "Using Concept Diagrams to Promote Understanding in Geometry." *Teaching Children Mathematics* 2 (November 1995): 184–89.

Tepper, Anita Benna. "A Journey through Geometry: Designing a City Park." *Teaching Children Mathematics* 5 (February 1999): 348–52.

University of Illinois at Chicago. *Math Trailblazers*. Dubuque, Iowa: Kendall/Hunt Publishing Co., 1998.

Vissa, Jeanne M. "Coordinate Graphing: Shaping a Sticky Situation." *Arithmetic Teacher* 35 (November 1987): 6–10.

Woodward, Ernest, and Rebecca Brown. "Polyhedrons and Three-Dimensional Geometry." *Arithmetic Teacher* 41 (April 1994): 451–58.